Halley's Comet

Halley's Comet

DONALD TATTERSFIELD

Basil Blackwell

© Donald Tattersfield 1984

First published 1984
Basil Blackwell Publisher Limited
108 Cowley Road, Oxford OX4 1JF, England

Basil Blackwell Inc.
432 Park Avenue South, Suite 1505
New York, NY 10016, USA

British Library Cataloguing in Publication Data

Tattersfield, Donald
Halley's comet.
1. Halley's comet
I. Title
523.6'4   QB723.H2
ISBN 0–631–13558–8 ✓

Also included in the Library of Congress Cataloguing in Publication lists.

Typeset by Katerprint Co. Ltd, Cowley, Oxford
Printed in Great Britain

The computer programs in this book have been tested and if properly keyed in to an appropriate computer will perform the tasks for which they are intended. The Publishers regret that they cannot enter into any correspondence concerning any aspect of the programs with any user.

Contents

Contents

Acknowledgements

The need for such a book as this, written especially for the layman and the scientifically minded reader, was suggested to me by my colleague James F. Harper. I have received much encouragement in the writing of it from my wife Margaret and my children Peter and Ann. My daughter has used her skills as a bilingual secretary in translating and typing letters to sources of information in France. The typing of the manuscript has been executed by my wife, and for this I am most grateful.

I am indebted to several authorities for the illustrations. The persons directly involved are chiefly Peter Hingley, Librarian of the Royal Astronomical Society, and the librarians of La Bibliothèque Nationale, Paris, the British Library, the University of Cambridge Library and of Trinity College, Cambridge. The Executive Secretary of the British Interplanetary Society, Mr L. J. Carter, has been most helpful especially in supplying details of the various space probes and the rockets on which they are to be launched.

Michael J. Hendrie, Director of the Comet Section of the British Astronomical Association, kindly supplied a relevant portion of the ephemeris of Halley's comet at this apparition prepared by Dr Donald K. Yeomans of the Jet Propulsion Laboratory, to whom we all owe much for his computational skill.

My thanks are due to Christopher Miller and Philip Jones who have checked my computer programs and made them more user-friendly.

The author and publisher gratefully acknowledge permission to reproduce the following illustrations: La Ville de Bayeaux: plate 10; City of Birmingham Museum and Art Gallery: plate 11; British Interplanetary Society: plate 18; California Institute of Technology, Palomar Observatory: plate 12; European Space Agency: plate 17; G. Y. Haig: plate 14; Dr Daniel A. Klinglesmith III: plate 19; Lick Observatory, University of California, Santa Cruz: plate 15; Mount Wilson and Las Campanas Observatories: plate 7; National Portrait Gallery, London : plate 1; Royal Astronomical Society: plates 3, 4, 5, 8, 9; Scala, Firenze: plate 6; the Master and Fellows of Trinity College, Cambridge: plates 2, 13.

I

Introduction

1 What is so special about Comet Halley?

Halley's comet was successfully found again on the morning of 16 October 1982, after an absence of about 72 years. Attempts to detect it were being made as long ago as 1977 with the 200-inch (5.1 metre) optical telescope at Mount Palomar, USA. The search was largely being conducted by professional astronomers because they have access to large telescopes with great light-gathering characteristics. Recently modern sensing devices, such as the charge-coupled detector (CCD) and associated equipment (see section 31), have made these telescopes even more efficient at recording the images of extremely faint objects. The search for new comets, and the rediscovery of comets which have been seen before, are very much occupations of amateur astronomers because large telescopes with an observer at the eyepiece are not suitable for sweeping the sky in search of new comets; although where there is a reliable forecast of the position of a returning comet a large telescope can be used to photograph that part of the sky. It is true that some new comets are discovered on photographic plates taken by the professionals for other purposes, and a few professional astronomers do spend some time in recovering old comets. Professor Elizabeth

Roemer, formerly of the United States Naval Observatory at Flagstaff, Arizona, and now at the Lunar and Planetary Laboratory University of Arizona, Tucson, USA, has a special reputation for this activity. The large telescopes are generally being used on more fundamental problems such as trying to solve the problem of the origin of our universe and expanding our knowledge of its evolution. Some studies in the infra-red part of the spectrum are undertaken, though these and studies in other parts of the spectrum (see section 23) are far better based in artificial satellites and with radio telescopes. Research on pulsars, quasars and black holes also takes up time on the large expensive telescopes. It therefore normally falls to the amateur astronomer with more modest equipment such as small telescopes and binoculars to spend many hours of the night and early morning searching the skies for new comets. These dedicated amateurs have an almost unbelievable knowledge of the positions of a very large number of stars in the sky, so they can more readily identify a newcomer, such as a comet.

No other comet has received such priority of attention from professional astronomers, so why is Comet Halley so special?

First, Halley's comet belongs to the family of comets known as *periodic*. This means that each travels in an elongated orbit round the sun forming a closed loop. So, with some exceptions, after being observed once from the Earth in that part of the orbit which takes them near to the Sun, and completing the remaining part of their orbit, they return to be visible on future occasions. Halley's comet is exceptional in that it has returned close to the Earth on no less than 29 previous occasions, having been observed at all but one of these appearances – that of 164 BC – since 239 BC. There is some evidence, however, that this comet has been seen on several more returns since 1140 BC, over a remarkable span of over 3000 years.

Second, whereas typically many periodic comets have orbits going out only as far as the orbit of the planet Jupiter and back in a period of about seven years, Halley's comet travels beyond the orbit of the farthermost planet of the solar system, Pluto, before returning. The former group of comets, approaching 100 in number, are ones which have passed sufficiently close to Jupiter on their way into, or out of, the solar system for their motion to have been materially disturbed by the gravitational effects of this massive planet. Their original orbits have consequently been changed to much smaller ones – see section 32.

Third, because of its very elongated orbit (see figure 5), the period of Halley's comet is about 76 years, though it has been as low as 74 and as high as 79. This is especially interesting because three score years and ten is still a reasonable average for the lifespan of man. Many people therefore will have the opportunity to observe this unusual comet twice in a lifetime, and no doubt have tales to tell about the last appearance, in 1910. As luck will have it, there was another bright comet early in 1910, the 'Daylight Comet', much brighter than Halley's, and it is likely that some of the tales will quite innocently be about the wrong comet.

Fourth, although reputable astronomers A. C. D. Crommelin and Professor Raymond Lyttleton consider that there must be tens of thousands of comets in the solar system which are theoretically observable, most of them are fainter than the faintest star visible with the naked eye (about magnitude 6; see section 8). Only a few are observed by dedicated astronomers, professional and amateur. Halley's comet has been seen at times which coincide with events of special historical significance, sometimes of a tragic nature. In AD 66 its appearance was taken as a warning to the Jews of the coming (AD 70) siege and capture of Jerusalem by the Romans under their emperor Vespasian. A little nearer our time, possibly the most notable appearance of Halley's comet was that which coincided with the Battle of Hastings in AD 1066. Its clear visibility at that time made it of sufficient importance to be represented pictorially in the Bayeux Tapestry which records the Norman conquest. Its appearance in the tapestry seems to be connected with disaster for the English, but its appearance in the sky could have been happily significant for Harold, Earl of Wessex, who had earlier been released from captivity by William the Bastard after a campaign in Brittany.

Last, we live in an age of technological and scientific achievement, which makes this return of Halley's comet one of intense interest to scientists. The instruments available today for scientific detection and measurement are as never before. In particular our ability to launch space probes with extreme accuracy over large distances has set afoot several projects to rendezvous with or fly by Comet Halley so that our understanding of the nature of comets in general, at present far from complete, can be greatly enhanced.

As Halley's comet approaches, to possibly within binocular range by 1985, interest in it will increase rapidly. A similar situation arose in 1973 when Comet Kohoutek took the imagination of the press. Because it was discovered by Dr Lubos Kohoutek at a distance of over 400 million miles from Earth, which is exceptional, it was assumed that this comet must be very large. Its calculated orbit would place it very favourably for observation, and a predicted magnitude (see section 22) of -12 (section 21) would make it very bright indeed, nearly as bright as the Moon. Daylight naked-eye visibility seemed certain. The reality was that Kohoutek's actual brightness fell far short of that predicted, well below naked-eye visibility, although Patrick Moore saw it from an aeroplane at magnitude about 3 and with a very short tail indeed. Such are the hazards of comet prediction (see section 22).

Nevertheless, as Halley's comet comes closer to the Earth and Sun, similar and more searching questions will naturally be posed by the layman, students in colleges and schools, and others. This book anticipates many of these questions and sets out to answer them in a straightforward way. Other relevant material, including some scientific background, is included to make the subject more intelligible and more interesting to those who wish to go deeper. These sections can be skipped without any loss by the general reader. Adequate cross-referencing and a small amount of repetition allow the reader to dip into the book at any place after reading this Introduction. A feature of the book which may be of particular interest to some readers is the inclusion of microcomputer programs.

2   Why is it called Halley's comet?

When an astronomer considers that he has seen a new comet, or has found on a photograph of a part of the sky an image which he thinks is a new comet, he uses the standard procedure for sending a report of his discovery to the International Astronomical Union's Central Bureau for Astronomical Telegrams (see appendix C). Steps are taken to confirm the original observations using other independent observers. If the object observed is confirmed, the new comet is then labelled with the year of discovery followed by the letter of the alphabet which indicates how many comets have previously been discovered in that year. If the observer is shown to be the first person to have discovered the new comet, it will be given his or her name. Thus, Comet Halley 1909c was the third comet to be discovered in 1909. The position of this comet had been predicted from observations made at its previous appearance in 1835, and so it did not carry the name of the person who first rediscovered it, on a photographic plate, in 1909 – thought to be Professor Max Wolf at Heidelberg.

After a period of about two years the comet will be given a more permanent label which indicates the year of its *perihelion* passage (when it is closest to the Sun) together with Roman numerals indicating the chronological order of that perihelion passage compared with other comets with perihelion passages in that year. Thus, Comet Halley was labelled 1910 II. If the comet is known to be periodic it will be given the name of the discoverer preceded by the letter P. So we have P/Halley 1910 II.

No such system existed in the time of Edmond Halley (1656–1742), nor did Halley actually discover the comet which bears his name. This will be clear from the number of appearances which the comet had already made before Halley was born. Nor did Halley rediscover it on its reappearance in September 1682, but what he did was truly remarkable.

Halley lived at the time of the publication of Isaac Newton's *Principia* (*Philosophiae Naturalis Principia Mathematica*). (Indeed, he not only encouraged Newton to publish this great work; he provided the funds to pay the printer and edited the work himself. He had some unusual rewards in addition to the profits from the first edition: free copies of a book on fishes published by the Royal Society.) Newton had put forward in *Principia* his theory of universal gravitation and had extended this to show that comets must travel in orbits of specific shapes (ellipses, parabolas or hyperbolas) (see section 4 and figure 1) round the Sun.

Halley took the observations of the comet of 1682 and, applying tedious computations (still tedious today unless one has a computer), deduced that its path must be elliptical. Newton himself had favoured parabolas for all comets. Certainly many comets when they are near the Sun move in elliptical paths which differ very little from parabolic paths (see figure 1). If indeed the path of the 1682 comet was elliptical, the comet must move round the closed

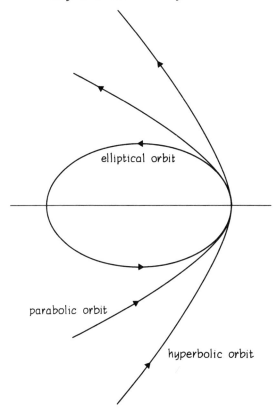

FIGURE I   *The shapes of the orbits of a comet moving under the gravitational influence of the Sun.*

loop of the ellipse and Halley predicted that it would reappear near the Sun in about 75 years. For the first time it was realized that not all comets are new; that some are reappearances of old ones. Halley's calculations would be put to the test in 1758. The comet was observed in December 1758, some say on Christmas Day, but Halley died in 1742 and never saw the reappearance of the comet which so clearly proved his calculations.

This comet is therefore called Halley's comet whatever number is assigned to it on each appearance. The comet was found again on 16 October 1982 and designated 1982i, being the ninth comet of 1982.

3 Who *was* Halley?

Edmond Halley (plate 1) lived from 8 November 1656, or 26 October 'by his own account', until 14 January 1742. This sort of slight discrepancy on dates runs through the literature on Halley, but they generally vary by less than a year. Halley and Newton were friends and each helped the other in his own way. Their characters and temperaments could scarcely have been more unlike.

Halley was born in Shoreditch, London. He attended St Paul's School before going up to Queen's College, Oxford, in 1673. It was here that he met John Flamsteed who was to become the first Astronomer Royal. One important problem of their time, which presents no difficulty at all in our day of artificial satellites, was the determination of one's position at sea, particularly longitude, and the post of Astronomer Royal was created in 1675 by King Charles II specifically to study this problem. John Flamsteed was a person whose later attitude and behaviour to both Halley and Newton were often antagonistic. The second Astronomer Royal was Edmond Halley himself, who held the post from 1721 until his death at the age of 85.

His visits to the Royal Greenwich Observatory, while he was a student at Oxford, stimulated his interest in astronomy. He left the university to go to St Helena where he spent a year cataloguing stars in the southern hemisphere. Flamsteed was carrying out a similar exercise for stars in the northern hemisphere, but the equipment was better at St Helena, and, at a latitude of 16° south of the equator, the island was the most southerly territory under British rule in

PLATE 1 *Edmond Halley, second Astronomer Royal.*

the south Atlantic. Halley's catalogue, published late in 1678, gave the coordinates of 341 stars in the southern hemisphere.

On his return to England in 1678 Halley was awarded his MA at Oxford, with some assistance from the King. In the same year he was elected a Fellow of the Royal Society of London, of which he became successively Clerk in 1686

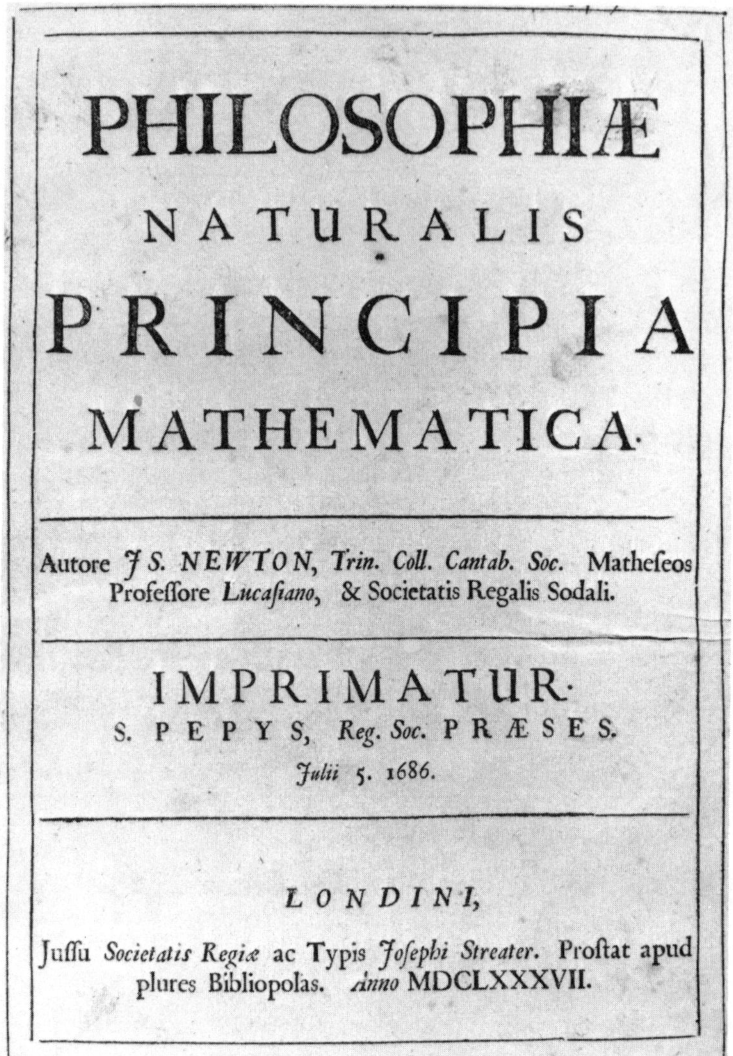

PHILOSOPHIÆ

NATURALIS

PRINCIPIA

MATHEMATICA.

Autore *J S. NEWTON, Trin. Coll. Cantab. Soc.* Matheseos
Professore *Lucasiano,* & Societatis Regalis Sodali.

IMPRIMATUR·
S. P E P Y S, *Reg. Soc.* P R Æ S E S.
Julii 5. 1686.

LONDINI,

Jussu *Societatis Regiæ* ac Typis *Josephi Streater.* Prostat apud
plures Bibliopolas. *Anno* MDCLXXXVII.

PLATE 2   *The title page of Newton's* Principia, *bearing the name of
Samuel Pepys.*

and Secretary in 1702, but he never became President; that honour fell to his friend Isaac Newton in 1703.

Newton's *Principia* (plate 2) might not have been published in Newton's day but for his friendship with Halley, for the former is known to have been rather lax at publishing his work. Halley not only paid for the printing, but he also did the proof reading and other editorial work. The original book sold to booksellers at 6 shillings unbound. We can gain some idea of the value put upon the *Principia* if we look at the accounts of the Royal Astronomical Society for 1836: 'J. Basire for drawing Telescopic Appearance of Halley's Comet 1836 . . . £8.10s' (plate 3).

In his long life Edmond Halley achieved distinction in several fields. Coupled with his diplomatic skills, his experience with the new telescopic sights on the telescope on St Helena led to his being to be selected to smooth out differences which had arisen between the respected astronomer Johannes Hevelius in Danzig and the Royal Society of London.

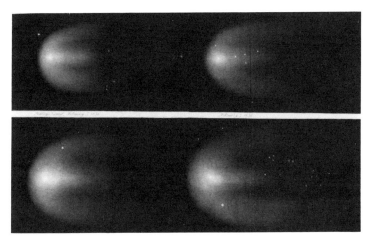

PLATE 3 *Drawings of Comet Halley 1836, prepared for the Royal Astronomical Society by J. Basire. The original drawings were made by Charles Piazzi Smyth at the Cape of Good Hope.*

Halley's visit to St Helena was fruitful in other ways too. On rare occasions the planet Mercury crosses the face of the Sun as viewed from the Earth. Halley observed one of these transits while on St Helena, and from it he was able to make a rough calculation of the *astronomical unit* (a.u.) (the radius of the Earth's orbit round the Sun – about 150×10^6 km). This experience must surely have influenced him in 1716 when he laid down a method of determining the astronomical unit using the transits of Venus which were predicted for 1761 and 1769. Of course Halley was not alive to see these transits, but the scale of the solar system was determined to an accuracy of about 5 per cent, a considerable improvement on the previous accuracy of 25 to 30 per cent.

The honour which did come to Edmond Halley in 1703

Cometarum Omnium hactenus rite Observatorum, Motuum in Orbe Parabolico Elementa Astronomica.

Cometæ Anni.	Nodus Ascend. gr.	Inclin. Orbitæ. gr.	Perihelion. in Orbe. gr.	Perihelion in Ecliptica gr.	Latitudo Perihelii gr.	Distantia Perihelii à Sole.	Log. dist. Perihelii à Sole.	Temp. æquat. Perihelii Londini. die. h.
1337	♊ 24. 21. 0	32. 11. 0	♉ 7. 59. 0	♉ 12. 45. 15	22. 40. 30 B	40660	9. 609230	Junii 2. 0. 25
1472	♑ 11. 46. 20	5. 20. 0	♉ 15. 33. 30	♉ 15. 40. 20	4. 25. 50 A	54273	9. 734584	Feb. 28. 22. 23
1531	♉ 19. 25. 0	17. 56. 0	♏ 1. 39. 0	♏ 0. 48. 15	17. 3. 05 B	56700	9. 753583	Aug. 24. 21. 18½
1532	♊ 20. 27. 0	32. 36. 0	♋ 21. 7. 0	♋ 16. 59. 40	15. 57. 00 B	50910	9. 700803	Oct. 19. 22. 12
1556	♍ 25. 42. 0	32. 6. 30	♍ 8. 50. 0	♑ 11. 0. 0	31. 10. 20 B	46390	9. 666424	Apr. 21. 20. 3
1577	♈ 25. 52. 0	74. 32. 45	♌ 9. 22. 0	♍ 7. 53. 0	09. 35. 20 A	18342	9. 263447	Oct. 26. 18. 45
1580	♈ 18. 57. 20	64. 40. 0	♌ 19. 5. 50	♎ 19. 17. 10	64. 40. 0 B	59528	9. 775450	Nov. 28. 15. 00
1585	♉ 7. 42. 30	6. 4. 0	♈ 8. 51. 0	♈ 8. 59. 10	2. 55. 25 A	109358	0. 038850	Sept. 27. 19. 20
1590	♍ 15. 30. 40	29. 40. 40	♏ 6. 54. 30	♏ 2. 55. 50	22. 45. 50 A	5,601	9. 700802	Jan. 29. 3. 45
1596	♒ 12. 12. 30	55. 12. 0	♏ 18. 16. 0	♏ 22. 44	35 54. 44 30 B	51293	9. 710058	Julii 31. 19. 55
1607	♉ 20. 21. 0	17. 2. 0	♒ 2. 16. 0	♒ 1. 29. 40	16. 10. 5 B	58680	9. 768490	Oct. 16. 3. 50
1618	♊ 16. 1. 0	37. 34. 0	♈ 2. 14. 0	♈ 6. 10. 00	35. 52. 0 A	37975	9. 579430	Oct. 29. 12. 23
1652	♊ 28. 10. 0	79. 28. 0	♈ 28. 18. 40	♈ 10. 41	35 58. 14. 0 A	84750	9. 928140	Nov. 2. 15. 40
1661	♊ 22. 30. 30	32. 35. 50	♌ 25. 58. 40	♍ 21. 37. 30	17. 17. 0 B	44851	9. 651772	Jan. 16. 23. 41
1664	♊ 21. 14. 0	21. 18. 30	♌ 10. 41. 25	♌ 8. 40. 35	16. 1. 50 A	1025753	0. 011044	Nov. 24. 11. 52
1665	♏ 18. 02. 0	76. 05. 0	♊ 11. 54. 30	♉ 24. 6. 35	23. 8. 0 B	10649	9. 027309	Apr. 14. 5. 15½
1672	♑ 27. 30. 30	83. 22. 10	♊ 16. 59. 30	♋ 9. 26. 00	69. 27. 40 B	69739	9. 843476	Feb. 20. 8. 37
1677	♏ 26. 49. 10	79. 03. 15	♊ 17. 37. 5	♋ 17. 16. 21. 0	75. 44. 10 B	28059	9. 448072	Apr. 26. 00. 37½
1680	♑ 2. 2. 0	60. 56. 0	♐ 22. 39. 30	♐ 27. 26. 50	8. 11. 10 A	00612	7. 787106	Dec. 8. 00. 6
1682	♉ 21. 16. 30	17. 56. 0	♒ 2. 52. 45	♒ 2. 0. 30	16. 59. 20 B	58328	9. 765877	Sept. 4. 07. 39
1683	♍ 23. 23. 0	83. 11. 0	♍ 25. 29. 30	♍ 10. 36. 55	82. 52. 00 B	56020	9. 748343	Julii 3. 2. 50
1684	♍ 28. 15. 0	65. 48. 40	♍ 28. 52. 0	♌ 15. 15. 25	26. 35. 20 A	96015	9. 982333	Maii 29. 10. 16
1686	♓ 20. 34. 40	31. 21. 40	♏ 17. 00. 30	♊ 16. 24. 00	31. 17. 35 B	32500	9. 511883	Sept. 6. 14. 33
1698	♒ 27. 44. 15	11. 46. 0	♍ 00. 51. 15	♑ 0. 47. 20	0. 38. 10 A	69129	9. 839660	Oct. 8. 16. 57

Hæc Tabula vix indiget explicatione, cum ex titulis satis pateat quid sibi velint Numeri. Distantiæ autem perihelii æstimantur in ejusmodi partibus quales media distantia Terræ à Sole habet centies millenas. Tabula

PLATE 4 *Halley's original table of the elements of the orbits of 24 comets given in* Synopsis Astronomiae Cometicae.

was his appointment as Savilian Professor of Geometry in Oxford in spite of the opposition of Flamsteed. As we shall see (section 4), conic sections figure prominently in the orbits of comets. Halley's major prediction that the comet of 1682 would return again in 1758 was based on his list of the elements of 24 comet orbits published in 1705 in his *Synopsis Astronomiae Cometicae* (see plate 4). His further interest in conic sections is demonstrated by his translation of the *Conics* of Apollonius in 1710.

Even before he became Astronomer Royal, Halley was interested in the problem of determining one's position at sea. He thought the answer lay in two things: a knowledge of the Moon's motion, and a knowledge of the direction of true north. The determination of the latter depends on a knowledge of the Earth's magnetic field. It was to chart the variation of this field that Halley undertook, in the years 1698–1700, a voyage in the Atlantic which was to take him as far south as the Falkland Islands. His method of charting is still used today. Halley began studying the Moon's motion, which is very complex and has a cycle extending over 18 years, in 1686 and continued his observations until he was 82 years of age.

Though the stars appear to be fixed in relation to one another, Halley compared the positions of the stars Sirius, Procyon and Arcturus given in the catalogues of Hipparchus and Ptolemy with positions given in contemporary catalogues. He came to the conclusion, published in a paper in 1708, that individual stars do have their own motion. This was no mean achievement, for the star with the greatest known movement of this kind, *proper* motion, moves through an angle of only half a degree in 180 years relative to the other stars.

Halley carried out a formidable variety of work, not only in astronomy but in diplomatic circles and on such mundane matters as producing mortality tables for insurance purposes. He married Mary Tooke in 1682, and they had two daughters and one son. His domestic life must have been very much interrupted by his extensive interests, but nevertheless he stayed happily married for 55 years.

Edmond Halley was an educated man in the true sense of the word, friendly and helpful to many. It is a pity that he did not live long enough to see his prediction on his comet fulfilled (he would have been 102 years old). He died at Greenwich in 1742, but he is remembered every 76 years. For this we may say Hallelujah!

II

Halley's Comet in 1985–1986

4 The motion of comets

Halley deduced that the comet of 1682 would travel in an elliptical path to return in 1758. Newton on the other hand favoured parabolic paths for all comet orbits near the Sun. If comets do maintain this parabolic motion they would not return, since the parabola is not a closed curve. (Some comets approach the Sun with a higher velocity so that their paths are along a hyperbola, which again is not a closed curve.) Strangely enough Kepler, who earlier had spent much of his life in coming to the conclusion that the planets revolve round the Sun in ellipses, considered that comets moved in straight lines 'with a motion not much different from a rectilinear one'.

All the curves mentioned above are known as *conics* because they are the shapes formed when an upright cone with a circular base is cut in different ways (see figure 2). The *eccentricity* (e) of a conic determines which type it is. Conics with an eccentricity of exactly 1 are parabolas. Those with $e > 1$ are hyperbolas and those with $e < 1$ ellipses. The eccentricity of an ellipse relates its length and width. A large eccentricity for an ellipse, such as 0.967 for Comet Halley, indicates a long narrow ellipse.

It is too simplistic to say that any comet will continue to

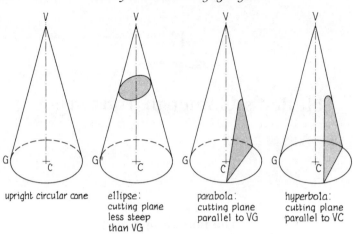

upright circular cone   ellipse:
cutting plane
less steep
than VG

parabola:
cutting plane
parallel to VG

hyperbola:
cutting plane
parallel to VC

FIGURE 2   *The generation of conic sections.*

travel precisely in the same ellipse, parabola or hyperbola as the one in which it is first observed. This would be true only if the sole body exerting its gravitational influence on the comet were the Sun. In fact, other massive bodies such as the planets, and particularly Saturn and Jupiter, have their influence on the comet's motion. They cause variations from the observed path, and in the lengthy prediction of the return of a comet these have to be taken into account. Halley appreciated this, because he knew that even the motion of Saturn is affected by Jupiter, and he made allowance for the effect of Jupiter in his prediction for the 1758 return of his comet.

In more recent times P. H. Cowell and A. C. D. Crommelin made a prediction for the return of Halley's comet in 1910 which proved to be incorrect by three days. One possible reason for this inaccuracy is that these two experienced computers took into account the effects of all the planets except Pluto, which was not discovered until 1930. More recently, in a search for a possible tenth planet, a planet X, in the solar system, Dr Joseph Brady of the

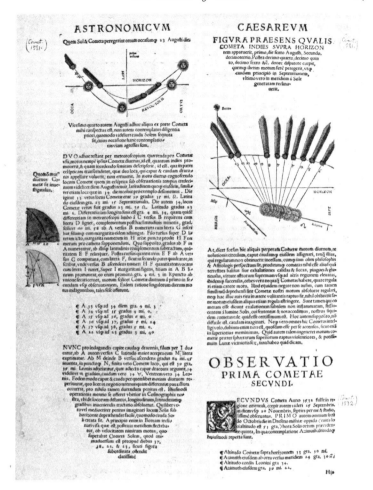

University of California has looked for variations in the orbit of Halley's comet which might reveal the existence of such a planet, but no tenth planet has been found. The space probes Pioneer 10 and 11, now on their way out of the solar system, are to be used in a similar exercise.

When a new comet is sighted, it is important to obtain quickly a predicted timetable, or *ephemeris*, of its motion in the immediate future so that track of it can be kept. It is standard practice to assume as a first approximation that the comet is travelling in a parabolic orbit. It is much easier and quicker to calculate the values, or elements, which define the path of the comet if this assumption is made.

We know that in 1705 Halley presented a paper entitled *Synopsis Astronomiae Cometicae* ('A Synopsis of the Astronomy of Comets') to the Royal Society of London, giving elements for 24 comets for which he had the most reliable information from historical records. In all cases he used Newton's new method for constructing a parabolic orbit. By comparing the elements of these orbits (which we also can do from plate 4), Halley was able to make the prediction:

> Now many things lead me to believe that the comet of the year 1531 observed by Apian [see plate 5] is the same as that which in the year 1607 was described by Kepler and Longomontanus, and which I myself saw and observed at its return in 1682. All the elements agree [see plate 4] except that there is an inequality in the times of revolution; but this is not so great that it cannot be attributed to physical causes. . . . I may, therefore, with confidence predict its return in the year 1758.

Halley was also keen that, if his predictions proved correct, all posterity would not fail to acknowledge that his discovery had first been made by an Englishman. Such national pride, and he was not yet the second Astronomer Royal!

5 Demonstrating the motion of a comet in its orbit around the Sun: computer programs

The following program in BASIC, called KEPLER, demonstrates the motion of a comet in an elliptical orbit round the Sun. All the calculations are completed before the display begins so the intervals between the positions of the comet are equal in real time. Note in particular the movement at each time interval as the comet recedes from the Sun.

The values suggested for a reasonable demonstration are given in table 1. The program is, however, flexible and some of the variables such as the semi-major axis and the eccentricity may be changed to demonstrate the effect of each on the orbit. When the eccentricity is near unity, as is the case with Halley's comet, the calculations can become unreliable at the points furthest from the Sun or if too large a time interval is chosen. The values for Halley's comet are also given in the table so that this may be demonstrated. However, the characteristics of the orbital motion of Comet Halley are demonstrated in the sample orbit.

TABLE 1

	Computer input	Sample orbit	Comet Halley
Semi-major axis	A	14 a.u.	17.94 a.u.
Eccentricity	EEC (E)	0.7	0.967
Time of perihelion passage	PP	1986.0	1986.11
Period	P	12 years	76 years
giving:			
Yearly motion	YM	360/12 = 30 degrees per year	360/76 = 4.74 degrees per year
Initial time	ST	1985.8	1985.8
Number of dates	U	24	74
Time interval	TI	0.5 years	1.0 years

Kepler's second law states that for a body, such as a comet, moving under the gravitational influence of the Sun only, the line joining the Sun to the body sweeps out equal areas in equal times. The programs for the BBC computer and the Spectrum demonstrate this visually and calculate the relevant areas for comparison. You might like to find out what Kepler's first and third laws are.

For explanatory notes on all the computer programs, see appendix A. For this program written for the ZX81, see appendix B.

BBC Model B

```
10 REM *KEPLER*
20 MODE 4
30 PRINT "ORBIT DEMONSTRATION PROGRAM"
40 PRINT
50 REPEAT INPUT "Semimajor axis (1-20)",A:UNTIL A>=1 AND A<=20
60 REPEAT INPUT "Eccentricity (0-0.95)",ECC:UNTIL ECC>=0 AND ECC<1
70 PRINT
80 INPUT "Time of perihelion passage (year)",PP
90 INPUT "Initial time (year)",ST
100 INPUT "Yearly motion (degrees/year)",YM
110 REPEAT INPUT "Number of dates (10-100)",U:UNTIL U>=10 AND U<=100
120 INPUT "Time interval (years, e.g. 0.5)",TI
130 PRINT "Please wait..."
140 DIM D(100):DIM R(100)
150 DIM K(100):DIM L(100)
160 DIM X(100):DIM Y(100)
170 LET Q=A*(1-ECC)
180 LET B=A*SQR(1-ECC*ECC)
190 LET THETAE=0
200 FOR I =1 TO U
210 LET THETAC=(YM*(ST+(I-1)*TI-PP))*PI/180
220 REPEAT
230 LET DIFF=THETAC-(THETAE-ECC*SIN THETAE)
240 LET THETAE=THETAE+DIFF/(1-ECC*COS THETAE)
250 UNTIL ABS(DIFF)<0.0001
260 REM CONVERT POLAR->CARTESIAN
270 LET X(I)=A*(COS THETAE-ECC)
280 LET Y(I)=B*SIN THETAE
290 NEXT I
300 CLG
310 VDU 19,1,3,0,0,0
320 VDU 19,0,4,0,0,0
330 VDU 23,241,0,0,0,0,0,2,7,2
340 VDU 23,244,0,0,0,0,0,0,3,3
350 VDU 5
360 MOVE 250,550
370 PRINT CHR$(241)
380 FOR I =1 TO U
390 LET K(I)=-30*X(I)
400 LET L(I)=30*Y(I)
410 LET R(I)=SQR(X(I)*X(I)+Y(I)*Y(I))
420 NEXT I
430 FOR I =1 TO U
440 MOVE 250+K(I),550+L(I)
450 PRINT CHR$(244)
460 MOVE 270+K(I),525+L(I)
470 IF R(I)<A THEN DRAW 270+K(I)-50*X(I)/R(I),525+L(I)+50*Y(I)/R(I)
480 FOR T=1 TO 1500:NEXT T
490 NEXT I
500 FOR I=1 TO U
510 MOVE 270,525
520 DRAW 270+K(I),525+L(I)
```

```
530 LET X(I)=270+K(I)
540 LET Y(I)=525+L(I)
550 FOR T=1 TO 1500:NEXT T
560 NEXT I
570 MOVE 0,30
580 INPUT "Press RETURN to calculate areas",a$
590 CLG:MODE 7:VDU 14
600 PRINT "AREAS"
610 FOR I=1 TO U-1
620 LET D(I)=270*(Y(I+1)-Y(I))-525*(X(I+1)-X(I))+(X(I+1)*Y(I)-X(I)*Y(I+1))
630 PRINT I;" ";I+1;TAB(15);INT D(I)
640 PRINT
650 NEXT I
660 VDU 4
670 END
```

Spectrum

```
10 REM "KEPLER"
20 PRINT "ORBIT DEMONSTRATION"
30 INPUT "Semimajor axis (1-20)",a
40 IF a<1 OR a>20 THEN GO TO 30
50 INPUT "Eccentricity (0-1)",e
60 IF e<0 OR e>0.95 THEN GO TO 50
70 INPUT "Time of perihelion passage (year)",pp
80 INPUT "Initial time (year)",st
90 INPUT "Yearly motion (degrees/year)",ym
100 INPUT "Number of dates (10-100)",u
110 IF u<10 or u>100 THEN GO TO 100
120 INPUT "Time interval (years e.g. 0.5)",ti
130 PRINT "Please wait..."
140 DIM k(100): DIM l(100): DIM r(100)
150 DIM x(100): DIM y(100)
160 LET q=a*(1-e)
170 LET b=a*SQR (1-e*e)
180 PAPER 1: INK 6
190 REM useful constants
200 LET RAD=PI/180
210 LET DEG=180/PI
220 LET thetae=0
230 FOR i=1 TO u
240 LET thetac=ym*(st+(i-1)*ti-pp)*RAD
250 REM next approximation:
260 LET diff=thetac-(thetae-e*SIN thetae)
270 LET thetae=thetae+diff/(1-e*COS thetae)
280 IF ABS (diff)>0.0001 THEN GO TO 250
290 REM convert polar to cartesian
300 LET x(i)=a*(COS thetae-e)
310 LET y(i)=b*SIN thetae
320 NEXT i
330 REM scale for screen
340 FOR i=1 TO u
350 LET k(i)=30-q+x(i)
360 LET l(i)=11+y(i)
370 LET r(i)=SQR(x(i)*x(i)+y(i)*y(i))
380 NEXT i
390 CLS
400 PLOT 210,83: DRAW 4,0: PLOT 212,81: DRAW 0,4
410 REM plot points
420 FOR i=1 TO u
430 PLOT 8*k(i),8*(21-l(i))
440 DRAW 1,1: DRAW -2,0: DRAW 0,-2: DRAW 2,0: DRAW 0,2
450 IF r(i)<a THEN DRAW (k(i)-212/8)*10/r(i),((21-l(i)-83/8))*10/r(i)
460 PAUSE 50
470 NEXT i
480 FOR i=1 TO u
490 PLOT 8*k(i),8*(21-l(i))
500 DRAW 212-8*k(i),83-8*(21-l(i))
510 PAUSE 50
520 NEXT i
530 PRINT "Press ENTER to display areas"
540 IF INKEY$="" THEN GO TO 540
550 CLS
560 PRINT "Areas:"
```

```
570 PRINT
580 FOR i=1 TO u-1
590 LET c=x(i)*y(i+1)-x(i+1)*y(i)
600 PRINT i;" to ";i+1;" = ";INT (c)
610 NEXT i
620 REM end of program
```

6   What Halley's comet will *not* do

First let it be made clear that Halley's comet will *not* flash across the sky so fast that you may miss it.

The constellations of stars, such as the Plough, or Great Bear, or Big Dipper, depending on where you live, hold their relative positions in the sky. Because of the rotation of the Earth on its axis from west to east through south the whole sky seems to move from east to west through south during the course of a night.

Halley's comet will be located in one of the constellations and will move with that constellation throughout the night. It will, however, be slow-moving as the days progress, passing from one constellation to the next. For instance, it will be seen in the constellation of Taurus the Bull during the latter part of September, the whole of October and the beginning of November 1985, passing into the constellation of Aries the Ram later in November 1985 and on into Pisces the Fishes in early December 1985. You might wonder why all the constellations mentioned so far are signs of the Zodiac, which astronomically are derived from the apparent position of the Sun in the sky as viewed over a period of one year. Halley's comet has an orbit which is inclined at only 18 degrees to that of the Earth and so for much of the time when the comet is nearest the Sun it will also appear in the zodiacal belt (see figure 3). It will not stay in this belt after it has passed successively through Aquarius the Water Bearer, Capricornus the Horned Goat, and Saggitarius the Archer (in late March 1986).

The objects which do flash across the sky and are seen only momentarily are *meteors*. They are debris in space which occasionally passes through the Earth's atmosphere

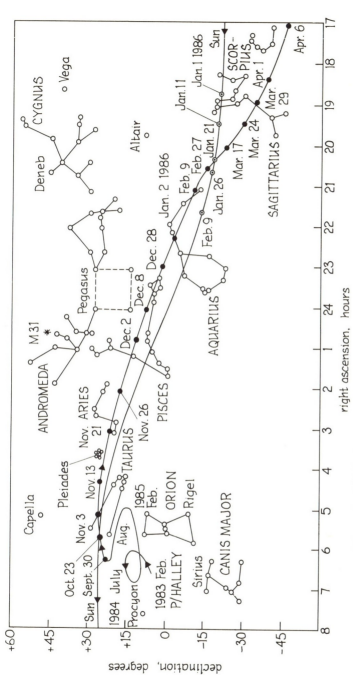

FIGURE 3   The path of Halley's comet in the sky, 1983–86. The position of the Sun in 1986 is also shown.

at large velocities of some 50 km/s. The friction between the debris and the Earth's atmosphere causes the debris to be heated up enough to become visible as meteors. Most of this material is very small and burns away before reaching the surface of the Earth.

Some of the debris in space comes from the material of comet tails (section 7) which streams away from the comets. The debris eventually becomes spread out along the orbit of its particular comet (see figure 4). It sometimes happens that each year the Earth, in its own orbit round the Sun, crosses the path of this debris. We then have a large increase in the number of meteors seen in a short time.

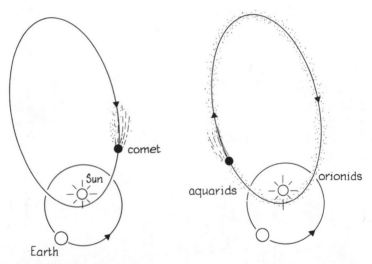

FIGURE 4 *Comet debris and the meteor showers originating from Halley's comet.*

Comet Halley, in its turn, has spread debris, and the meteor showers seen each year in April/May, the Aquarids, and again in October, the Orionids, arise from this. A rate of meteors of the order of 20 per hour is common. Because of a perspective effect, meteors in a particular shower appear to come from a particular part of the sky, known as the radiant. The location of the radiant of the Aquarids is, for example, in the constellation of Aquarius from which the shower takes its name.

Revolving around the Earth are a very large number of artificial satellites which have been launched from the Earth since the first Sputnik in 1957. Many of them travel once round the Earth in about 90 minutes and some of them can be seen easily with the naked eye. They are seen by the sunlight reflected from them high above the Earth's atmosphere, and look very much like a slowly moving star, crossing the sky in a few minutes. The author recently saw nine satellites in half an hour on one evening when looking for meteors of the Perseid shower. None of these satellites can be mistaken for Halley's comet.

In days past it was often thought by the superstitious that the appearance of a comet indicated a disaster of some kind. This must have referred to bright comets because many comets pass unnoticed. We have already mentioned the appearance of Halley's comet at the time of the Battle of Hastings – disaster for the English but not for the Normans. The return of Halley's comet in AD 66 when the Jewish war began was taken to be a sign that Jerusalem would fall to the Romans, even though this did not happen until AD 70. The 1456 appearance was associated by some with the fall of Constantinople, though Ottoman Sultan Mehemed II captured this city on 29 May 1453, three years earlier. By 1456 the Turks were at the gates of Belgrade, so you can *fit* dates to comet appearances.

Equally the appearance of a bright comet could coincide with a happy event. The painter Giotto di Bondone clearly associated the appearance of a comet with the birth of Jesus, though the comet he so realistically painted was Halley's comet on its return in 1301 (plate 6). The nature of the 'star' of Bethlehem continues to be discussed, but the likelihood of its being a bright comet has been discounted.

PLATE 6   *Giotto di Bondone's* Adoration of the Magi, *including Halley's comet 1301 as a detail.*

It is thought that Jesus was born between 7 BC and 5 BC, and the nearest return of Halley's comet took place in 11 BC.

The bright comet of 1680, not Halley's but sometimes called 'Newton's comet' because it assisted him in his conclusions set out in the *Principia* on the nature of comets, caused an international sensation. Many important personages in France were alarmed, thinking that this comet had come to warn them of their deaths. Some said that it preceded a return of the biblical flood, 'as they say that water is always announced by fire, unless M. Cassini should take the trouble to refute it'. Nevertheless a number of people made wills in favour of monasteries. Halley himself thought that the original flood (Genesis 7) might have been caused by the impact of a comet with the Earth or at least a close approach.

Even in 1910 there was great alarm on the return of Halley's comet. Increased scientific ability in spectroscopy (see section 23) had shown that comets contain a wide range of atoms and molecules (see section 18). One of the complex molecules found in comets is cyanogen (CN), a colourless, poisonous gas whose smell reminds some people of peaches. It burns with a violet-coloured flame. It was predicted that the Earth would pass through the tail of Halley's comet in 1910, and the *New York Times* carried both comforting and alarmist headlines (see section 27). It is reported that some people committed suicide in anticipation of possible distressing effects. Comet pills were marketed and sold well. Barrels of water were prepared, so that one's safety could be assured by immersing oneself up to the neck – even though cyanogen dissolves readily in water.

Recently some quite eminent astronomers, notably Sir Fred Hoyle and Chandra Wickramasinghe, have published well-researched but controversial works with the theme that epidemics are borne by comets (section 28). Influenza epidemics are quoted widely in this context.

Comet Halley at the coming return should not provoke such extreme reactions as the earlier ones described above.

7   What *will* Halley's comet look like?

The photographs of Halley's comet at its 1910 appearance (plates 7, 8, 9) show the comet as a diffuse object with a bright head and a fainter, extended tail. A camera attached to a telescope has the advantages over the naked eye that the telescope itself can collect more light and an extended exposure of the film can continue to add even more light to

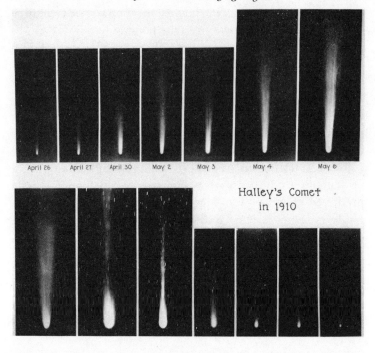

April 26   April 27   April 30   May 2   May 3   May 4   May 6

Halley's Comet
in 1910

PLATE 7   *Fourteen views of Halley's comet taken between 26 April and*
11 June 1910.

the image being formed. There is a tendency for the head of
the comet to be over-exposed when the exposure is unduly
long so that the detail of the fainter tail may be obtained.
The photographs may therefore give an exaggerated
impression to the naked-eye observer.

When the comet comes within naked-eye visibility it will
appear as a hazy patch of light with no well-defined
features. Observation over a few nights will show that this
patch of light has moved its position relative to the stars
near to it. As the comet approaches nearer to the Sun the
head, or *coma*, of the comet may take on the appearance of a
bright disc within a diffuse area surrounding it. The size of
the coma cannot be predicted; it changes as the comet

PLATE 8   *Halley's comet photographed at Helwan, Egypt, on 2 June 1910.*

PLATE 9   *The bow front of Halley's comet photographed at Helwan Observatory, Egypt, on 21 May 1910.*

approaches the Sun, and can be vastly different from comet to comet. During its return in 1910 Halley's comet had a coma of about 20 000 km diameter when it was barely detectable at a distance of 3 astronomical units (a.u.) from the Sun (between the orbits of Mars and Jupiter). At 2 a.u. it had grown to about 300 000 km, but it had *shrunk* to a diameter of about 200 000 km when at a distance of 0.6 a.u. from the Sun (within the orbit of the Earth). This is approximately one-quarter of the angular diameter of the full Moon. After passing its perihelion point the head of Halley's comet in 1910 once again expanded to similar dimensions at similar distances.

As Halley's comet approaches the Sun it is possible that the central part of the coma will appear to condense into a brighter, almost starlike point of light. This is often referred to as the *nucleus* of the comet but it is not known whether the nucleus actually exists or whether it is merely an optical effect (section 11).

Nor can the length of the comet's tail be predicted. The tail consists of dust and gas evaporated from the coma. The dust continues as separate solid particles each moving in its own solar orbit (section 18). The gas is driven by solar radiation pressure and the solar wind in a direction away from the Sun so the tail will point generally away from the Sun, a fact noted by Peter Apianus as long ago as 1531. The cometary diagrams (see plate 5) in his beautifully illustrated book *Astronomicum Caesareum* show this clearly, though most of the diagrams show the Sun below the horizon so the fact would not then be so obvious. Wên Hsien Thung Khao also noted this much earlier. He wrote of Halley's comet in AD 837 'When it appears in the morning the tail points to the west; when it appears in the evening to the east. This is a constant rule.' The solar wind consists of a stream of electrons and protons (see section 19) known as *plasma* emanating from the Sun. Its velocity has been measured by recent space probes as some 400 km/s at a distance of 1 a.u. from the Sun. The tail is not simply material left behind as the comet moves forward in its orbit. Indeed, when the comet is moving away from the Sun after its passage round it, the tail will precede the coma. The gaseous part of the tail may split from the main stream. The tail will become longer as the comet gets nearer to the Sun. On the first return in 1758 predicted by Halley, the comet at one time, May 1759, had a tail which spread through an arc of 47 degrees. In 1910 Halley's comet had a very long tail reaching over an arc of 140 degrees.

With the aid of binoculars or a telescope more detail may be seen, particularly in the structure of the coma. The central region may show up as a very small bright nucleus, while the diffuse area of the comet may take up a round-nosed shape on the side nearest the Sun (see plates 9, 3). The tail may seem turbulent and may have more than one stream. It is such a nebulous object that stars may be seen through it.

8   How bright will it be?

On many of its previous appearances Halley's comet was bright enough to be easily visible at night with the unaided eye. This is evident from the art and literature referring to its appearance before the invention of the telescope, early in the seventeenth century. Probably the most publicized appearance of Halley's comet is that of 1066 which is recorded in the Bayeux Tapestry (plate 10) with the text *Isti mirant stella* ('These men marvel at the star') by the side of a representation of the comet, while onlookers point it out. The painter Giotto will become better known world-wide because the European spacecraft which will be sent to fly by Comet Halley in 1986 has been given his name (section 30). Giotto's painting in the fresco in Padua (plate 6) includes a realistic detail of the 1301 appearance of this

PLATE 10   *Halley's comet 1066 illustrated in the Bayeux Tapestry.*

comet. In 1456 Paolo Toscanelli, a contemporary in Florence of Leonardo da Vinci, described Halley's comet as big and terrible, with a head as large as the eye of an ox and a long tail like that of a peacock. Toscanelli, 'Paul the Physician', practised medicine as well as astronomy so he should have had a good idea of the size of an eye. A more recent portrayal of Comet Halley's appearance, in 1759, is depicted on the Wedgwood medallion of Isaac Newton issued around 1780 (plate 11). The most recent appearance, in 1910 (see plates 7, 8) is so well documented that we can be certain the comet could be seen without difficulty in the early morning sky and, later, in the early evening sky.

Astronomers have a scientific way of describing the brightness of heavenly bodies (see section 21) which originated with the ancient Greek Hipparchus (*c.* 140 BC), who catalogued upwards of 850 stars by latitude and longitude on the celestial sphere. If we take the faintest stars which most people can see on a good starry night, astronomers say that these are of magnitude 6. Magnitude here does not refer to size but to apparent brightness. Again, if we take the brightest stars which we can see, leaving out the odd exceptionally bright ones like Sirius, the Dog Star, a magnitude of 1 is allotted to these. Thus the greater the magnitude the fainter the star. The eye can divide this range into units because it has *logarithmic sensitivity* (see section 21). On this scale the Pole Star has a magnitude of 2.12.

When Comet Halley was found again, or *recovered*, in 1982 it had a magnitude of about 24, which is very faint indeed and at about the limit of the most sensitive detecting device currently available (section 31).

The accurate prediction of the brightness of comets, particularly those with a long period like P/Halley, is very difficult. As the comet passes close to the Sun it loses volatile material, the amount of which is uncertain. Predictions of brightness at its next return, or *apparition*, are based

PLATE 11   *The appearance of Halley's comet in 1759 as depicted in the Wedgwood medallion of Isaac Newton issued around 1780.*

on a formula which carries two unknown terms. The value of these used in the prediction depends very much on experience of the behaviour of the comet at previous apparitions (see section 22). When the comet has been recovered and a number of observations of its magnitude have been made the predictions for future dates become more certain.

Nevertheless, well before Halley's comet was recovered Dr Donald Yeomans, of the Jet Propulsion Laboratory in the USA, made predictions of the brightness of Comet Halley from 1981 to 1987. Its brightness on recovery was not very different from 23.2 predicted by Dr Yeomans. We may reasonably expect that his other predictions of brightness will be correct, but the method of calculation of the brightness of a very distant comet is different from that when the comet is closer to the Sun. Halley's comet has in the past behaved differently before it is closest to the Sun, at perihelion, than when it is receding from the Sun.

By September 1985 the magnitude of P/Halley is predicted to be about 12. From that time onwards the brightness will increase slowly to about 6 in December 1985 and to between 3 and 6 in February 1986. For a few dates on each side of the perihelion passage on 9 February 1986 the comet will be on the side of the Sun opposite that of the Earth, so it will not be visible to us. On its reappearance from behind the Sun it will begin to fade to about magnitude 8 in June 1986, and to still higher magnitudes thereafter. The magnitudes referred to in this paragraph are the total magnitudes, that is for head plus tail, as opposed to the magnitude of the head only (see section 7). The total magnitude is given in figure 12 for dates between the beginning of November 1985 and the end of May 1986.

9 How fast is it moving?

When we first look at Halley's comet it will appear to be stationary in the sky against the background of the stars. We shall only notice differences in its position over periods of hours or even days. However, its true speed in its orbital path will be about 54 km per *second* when it is nearest the Sun. This is the astonishing speed of 121 500 miles per hour.

Halley's comet travels round its large orbit in 76 years, but its speed in its orbital path is far from constant. As it recedes from the Sun its speed drops, to a value of only about 1 km/s when the comet is furthest from the Sun, and when, of course, we cannot observe it.

Figure 5 shows the position of Comet Halley for various times in its present orbit, that is since its last appearance in 1910. Notice in particular that Comet Halley stayed beyond the orbit of Neptune from 1929 to 1967, almost half of the time for a full orbit. Yet in 13 years it will have travelled almost the half of the orbit nearest the Sun, from A to B.

The computer programs on pages 20–2 plot the position of a comet at equal time intervals, demonstrating that a comet stays for a long time near *aphelion*, the point in its orbit furthest from the Sun. It is not practical to use Halley's comet for this demonstration because of its long period and also because of its highly elongated orbit, but the positions shown are true for the comet used and they are similar for Comet Halley.
they are similar for Comet Halley.

The speed v of Halley's comet in its orbital path depends only upon the distance r of the comet from the Sun. The formula (see section 24) for the calculation of Comet Halley's orbital speed is simple:

$$v = 29.7 \times \sqrt{\left(\frac{2}{r} - 0.0557 \right)} \ \text{km/s}$$

where the value of r to be substituted is the distance of the comet from the Sun in astronomical units at the position for which we require the speed. As an example, the speed of the *comet* at encounter with the Giotto space probe at a distance of 0.89 a.u. from the Sun (see table 3) will be:

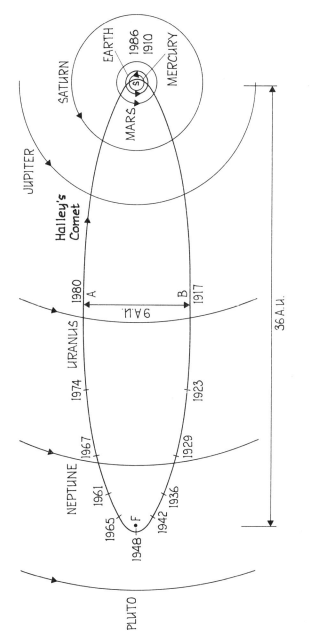

FIGURE 5   *The orbit of Halley's comet in relation to those of the planets.*

$$v = 29.7 \times \sqrt{\left(\frac{2}{0.89} - 0.0557\right)} = 29 \times 2.191 = 43 \text{ km/s.}$$

Taken with the velocity of Giotto itself, the velocity of approach will be nearly 70 km/s. This high speed makes the bumper shield on Giotto essential. This will be similar in form to that shown in figure 26. The shield will be an annulus of nearly 2 metres outside diameter surrounding the central main rocket motor of Giotto.

A dust particle of mass only one-tenth of a gram travelling at 70 km/s has as much energy as a car of mass 1500 lb travelling at 60 mph. Such a small mass striking solid aluminium would cause a crater about 2 cm deep, but the design of the bumper shield is such that masses from one-millionth of a gram up to one-tenth of a gram will be vaporized on striking the front sheet. Their impact shock will be absorbed over the large area of the high-resilience rear shield. Dust impact sensors will be placed on the front sheet. The mass distribution is to be studied down to 10^{-4} micrograms extended down to 10^{-7} micrograms and possibly further to 10^{-11} micrograms.

The high velocity of encounter means that the chemistry of the coma must be studied in the 10 seconds or so between Giotto entering the coma and its possible destruction near the nucleus. Apart from its nucleus a comet is a very tenuous object, so with luck Giotto might penetrate further than this, thereby increasing the very short time for its chemistry studies.

To set against the disadvantages of the high velocity of encounter, it has been calculated that a hydrogen atom moving with the comet at this speed has more energy than the atoms from outgassing of the probe itself and so can be distinguished from it. The high energy of other atoms will make it possible for the mass spectrometer to operate successfully.

10   When can we see it?

Forecasts about Comet Halley at this return are that it will not be so spectacular as at some previous apparitions. According to Donald Yeomans the total magnitude of the comet will be about 13 in mid-September 1985, becoming progressively brighter up to the time when it is nearest the Sun on 9 February 1986, when it should have a magnitude of about 3. It will then diminish in brightness to about magnitude 11 by mid-September 1986. Thus it should be possible for you to see Halley's comet during that span of about a year, depending on the factors mentioned in section 13.

The *elongation* of the comet is the angle made at the observer by the lines drawn from the Sun to the observer and from the observer to the comet. Whatever your latitude, the elongation of the comet will be less than 10 degrees between 2 February 1986 and 11 February 1986, so it will be difficult, if not impossible, to see the comet between these dates, because of the glare of the Sun. Indeed, Dr Porter used to consider that it was not useful to produce a timetable for a comet for the period when the elongation is less than 30 degrees.

Figures 13–18 will tell you when Halley's comet is visible in a dark sky at places at or near your latitude. An explanation of how to use these diagrams is given in section 14. The approximate coverage is:

60°N	Sweden
50°N	UK, France, Germany, Canada
40°N	Italy, Spain, Portugal, Northern USA, Japan
30°N	Southern USA
30°S	South Africa, Australia
40°S	New Zealand

11   Where will Comet Halley be?
Computer programs to determine its altitude and bearing

During the whole of 1984 the comet will be almost stationary a little above the two bright stars Procyon (magnitude 0.5) in the constellation of Canis Minor, the Little Dog, and Betelgeuse (magnitude variable 0.5–1.1) in the constellation of Orion, the Hunter. Over the course of the year, Comet Halley will appear to move slowly in a small loop against the backcloth of stars of the Milky Way. The magnitude of the comet during this time will be greater than 13, so it is most unlikely that you will be able to see it in this period unless you have an 8-inch reflecting telescope or one larger.

This loop, and the other smaller loops in which it moved earlier than 1984, are due to a combination of two motions. The comet is, of course, moving in its own orbit round the Sun; in addition the platform from which we are observing, the Earth, is also moving in its own orbit round the Sun so that the comet will be seen from different positions throughout the year and will therefore appear to be displaced.

As Comet Halley approaches the Sun and the Earth, its velocity in its own orbit will increase, from about 23 km/s (51 750 mph) in September 1985 to 38 km/s (85 000 mph) at the end of November 1985 when the comet will be about as far from the Sun as the planet Mars. The second effect, that of *parallax*, becomes of less importance and so from September 1985, when the magnitude of the comet will be less than 13, Halley's comet will appear to move continually westwards in the sky, relative to the background of the stars.

Both of these features can be seen in figure 3. The axes of this map are *declination* vertically and *right ascension* horizontally. These are terms associated with the way in which astronomers describe the position of a body in the sky and also the way in which they may direct their telescopes. Roughly, we may consider them to be equivalent to latitude and longitude respectively on Earth. If we stand at the centre of the Earth and project the Earth's equator on to the the sky, this is the *celestial equator* from which declination is measured in the same way that latitude is measured on earth. In astronomy longitude on Earth is now measured east from the line through the poles and Greenwich, England; right ascension is measured eastwards from the point in the sky where the Sun is on 21 March, the *vernal equinox*. The point is known as the First Point of Aries because it used to be in the constellation of Aries.

Although the axes of the map in figure 3 have been shown vertical and horizontal they should not be confused with *altitude*, degrees above the observer's horizon, and *azimuth*, the bearing in degrees measured eastwards from north.

A more useful representation for the layman is given in figures 6–9. These maps locate the position of Comet Halley at five-day intervals by its relation to the observer's horizon. The axes on these maps are altitude vertically and azimuth horizontally. The vertical scale is shortened towards the top to give the effect of looking at the inside of the celestial sphere. This method of representation does however limit the number of positions which can be shown reasonably, but you can easily interpolate for dates not specifically shown. Use the positions for the latitude nearest your own. The positions given are those when darkness has just fallen and also, if the comet is visible just before darkness ends, in the morning. Comet Halley will be visible all night for latitudes 30°S and 40°S from 8 April to 20 April 1986. Where no curve is given for early morning the comet will not be visible at the beginning of twilight.

The reader with access to a microcomputer can use the

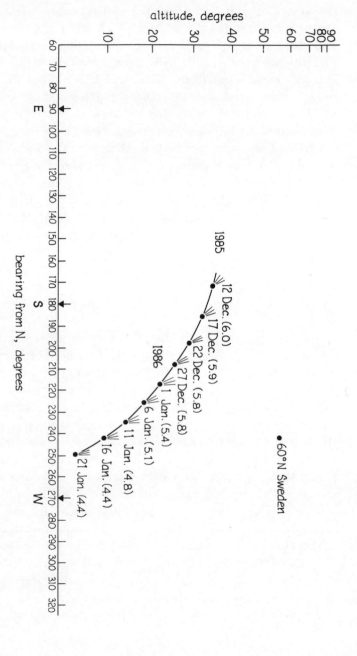

FIGURE 6 Comet Halley at evening twilight from latitude 60°N.

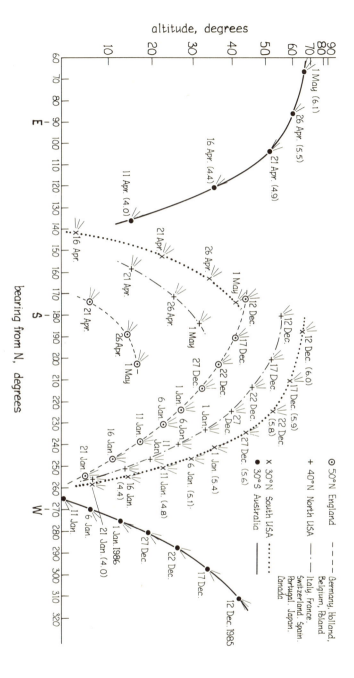

FIGURE 7 *Comet Halley at evening twilight from latitudes 50°, 40°, 30°N and 30°S.*

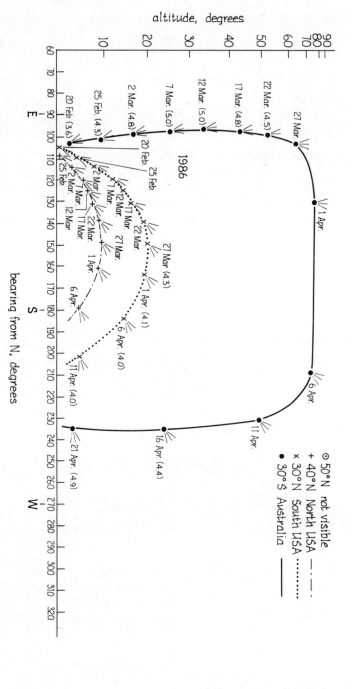

FIGURE 8   Comet Halley at morning twilight from latitudes 50°, 40°, 30°N and 30°S.

altitude, degrees

bearing from N, degrees

⊙ 50°N   not visible
+ 40°N   North USA — · — · —
× 30°N   South USA · · · · · · ·
● 30°S   Australia ————

1986

20 Feb. (3.6)
25 Feb. (4.3)
2 Mar. (4.8)
7 Mar. (5.0)
12 Mar. (5.0)
17 Mar. (4.8)
22 Mar. (4.5)
27 Mar.
1 Apr.
6 Apr.
11 Apr.
16 Apr. (4.4)
21 Apr. (4.9)

20 Feb.
25 Feb.
2 Mar.
7 Mar.
12 Mar.
17 Mar.
22 Mar.
27 Mar.
1 Apr.

27 Mar. (4.3)
1 Apr. (4.1)
6 Apr. (4.0)
11 Apr. (4.0)

E   S   W

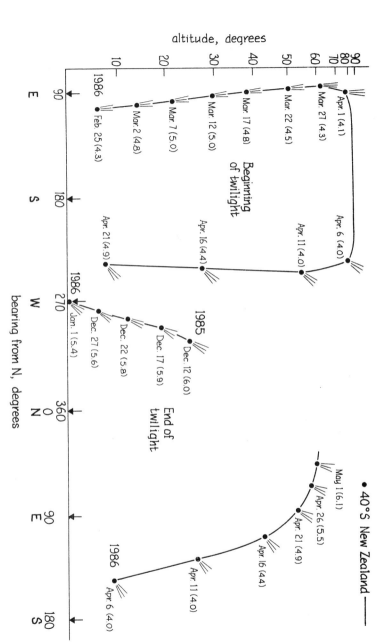

FIGURE 9   Comet Halley at evening twilight and at morning twilight from latitude 40°S.

program ALTAZ given at the end of this section to obtain the coordinates of the comet at any time in 1985 or 1986 as observed at any latitude or longitude. For explanatory notes on all the computer programs, see appendix A.

The maps will be useful for your first sighting of Halley's comet, from which time you should have little difficulty in following its progress night by night until it passes below your visible horizon. The predicted brightness of Comet Halley, its magnitude, is shown in brackets after each date shown; or it can be read from figure 12.

For a period on each side of 9 February 1986 the comet will not be visible since it will be passing behind the Sun, whose position in the sky around that date is also shown in figure 3.

BBC Model B

```
10 REM "ALTAZ"
20 MODE 4
30 PRINT
40 PRINT
50 PRINT "Altitude and Azimuth of Halley's"
60 PRINT "Comet and the Sun between"
70 PRINT "1st January 1985 and 31 December 1986"
80 DIM A(5):DIM B(5):DIM C(5):DIM D(5):DIM E(5):DIM F(5):DIM G(5)
90 DIM H(5):DIM I(5):DIM K(5):DIM L(5):DIM M(12):DIM N(5)
100 DIM P(5):DIM Q(5):DIM R(5):DIM S(5):DIM T(5)
110 DIM U(5):DIM V(5):DIM W(5):DIM X(5):DIM Y(5):DIM Z(5)
120 REM *CUMULATIVE MONTH DAYS*
130 FOR I = 1 TO 12
140 READ M(I)
150 NEXT I
160 DATA 0,31,59,90,120,151,181,212,243,273,304,334
170 REM *INPUT CONSTANTS COMET*
180 E(1)=0.967:W(1)=170.011:A(1)=17.94
190 Z(1)=58.1530:I(1)=162.238
200 REM *INPUT CONSTANTS SUN*
210 W(2)=282.510396:G(2)=279.041470:O=RAD (23.44)
220 REM *INPUT CONSTANTS EARTH*
230 E(3)=0.01672:W(3)=102.51044:G(3)=99.53431
240 REM *INPUT OTHER CONSTANTS*
250 A=0.065709:C=1.002743
260 PRINT " "
270 REPEAT PRINT"Enter latitude of observer in degrees"
280 INPUT "(0-90, + for North, - for South)",L:UNTIL L>=-90 AND L<=90
290 L=RAD (L)
300 PRINT
310 REPEAT INPUT "Enter year (1985 or 1986)",Y$
320 UNTIL Y$="1985" OR Y$="1986"
330 IF Y$="1985" THEN LET B=17.359625
340 IF Y$="1986" THEN LET B=17.375540
350 PRINT
360 REPEAT PRINT "Enter month":INPUT "(1 for January...12 for December)",N
370 UNTIL N>=1 AND N<=12
380 PRINT
390 REPEAT INPUT "Enter day of month (1-31)",D
400 UNTIL D>=1 AND D<=31
410 PRINT
```

```
420 REPEAT INPUT "Enter hour (0-23)",H
430 UNTIL H>=0 AND H<=23
440 PRINT
450 REPEAT  INPUT "Enter minute (0-59)",M
460 UNTIL M>=0 AND M<=59
470 REM *FOR SUN*
480 IF Y$="1985" THEN LET X=3653+M(N)+D+(H+M/60)/24
490 IF Y$="1986" THEN LET X=4018+M(N)+D+(H+M/60)/24
500 V=360*X/365.25
510 IF V>=0 AND V<=360 THEN GOTO 550
520 V=V-360
530 IF V>=0 AND V<=360 THEN GOTO 550
540 GOTO 520
550 Y=V+G(2)-W(2)
560 IF Y<0 THEN LET Y=Y+360
570 Y=RAD (Y)
580 E=(360*E(3))*SIN Y)/PI
590 L(2)=V+E+G(2)
600 IF L(2)>360 THEN LET L(2)=L(2)-360
610 L(2)=RAD (L(2))
620 D(2)=ASN(SIN O*SIN L(2))
630 R(2)=ACS(COS L(2)/COS D(2))
640 IF COS O*SIN L(2)>0 THEN LET R(2)=R(2)
650 IF COS O*SIN L(2)<0 THEN LET R(2)=2*PI-R(2)
660 PRINT
670 PRINT "DATA FOR THE SUN:"
680 PRINT
690 PRINT "Right Ascension:";TAB(20);DEG(R(2));" ";"degrees"
700 PRINT "Declination:";TAB(20);DEG(D(2));" ";"degrees"
710 X(5)=H+M/60
720 N(5)=M(N)+D
730 P(5)=N(5)*A-B+X(5)*C
740 IF P(5)>24 THEN LET P(5)=P(5)-24
750 IF P(5)<0 THEN LET P(5)=P(5)+24
760 H(2)=RAD (P(5)*15)-R(2)
770 H(5)=H(2):D(5)=D(2)
780 GOSUB 1760
790 V(2)=V(5):U(2)=U(5):K(2)=V(2):O(2)=U(2)
800 PRINT
810 PRINT "Azimuth:";TAB(20);DEG(V(2));" ";"degrees"
820 PRINT "Altitude:";TAB(20);DEG(U(2));" ";"degrees"
830 PRINT
840 PRINT
850 IF DEG(U(2))>-18 THEN PRINT "Sky not dark"
860 REM *FOR COMET*
870 F=360*(X-4058.66)/(365.25*76)
880 IF F<0 THEN LET F=F+360
890 IF F>=0 AND F<=360 THEN GOTO 930
900 F=F-360
910 IF F>=0 AND F<=360 THEN GOTO 930
920 GOTO 900
930 F=RAD(F)
940 Q=F
950 F(1)=Q-E(1)*SIN Q
960 T=ABS(F-F(1))
970 IF T<0.0001 THEN GOTO 1010
980 C(1)=(F-F(1))/(1-E(1)*COS Q)
990 Q=Q+C(1)
1000 GOTO 950
1010 G=SQR((1+E(1))/(1-E(1)))*TAN (Q/2)
1020 N(1)=ATN G:N(1)=2*N(1)
1030 W(1)=RAD(W(1)):Z(1)=RAD(Z(1)):I(1)=RAD(I(1))
1040 L(1)=N(1)+W(1)
1050 R=(A(1)*(1-E(1)*E(1)))/(1+E(1)*COS N(1))
1060 G(1)=L(1)-Z(1)
1070 P(1)=ASN(SIN G(1)*SIN I(1))
1080 L(4)=ATN(TAN G(1)*COS I(1))+Z(1)
1090 IF G(1)>=PI/2 AND G(1)<=3*PI/2 THEN LET L(4)=L(4)+PI
1100 IF L(4)<0 THEN LET L(4)=L(4)+2*PI
1110 S=R*COS P(1)
1120 REM *FOR EARTH*
1130 V=(360*X)/(365.25*1.00004)
1140 IF V>=0 AND V<=360 THEN GOTO 1180
1150 V=V-360
1160 IF V>=0 AND V<=360 THEN GOTO 1180
1170 GOTO 1150
1180 Y(3)=V+G(3)-W(3)
```

```
1190 Y(3)=RAD(Y(3))
1200 L(3)=V+(360*E(3)*SIN Y(3))/PI+G(3)
1210 IF L(3)<0 THEN LET L(3)=L(3)+360
1220 IF L(3)>360 THEN LET L(3)=L(3)-360
1230 L(3)=RAD (L(3))
1240 W(3)=RAD (W(3))
1250 N(3)=L(3)-W(3)
1260 S(3)=(1-E(3)*E(3))/(1+E(3)*COS N(3))
1270 LET DIFF=L(3)-L(4)
1280 IF S<S(3) THEN LET L(5)=PI+L(3)+ATN((S*SIN(DIFF))/(S(3)-S*COS(DIFF)))
1290 IF S>S(3) THEN LET L(5)=ATN((S(3)*SIN(-DIFF))/(S-S(3)*COS(-DIFF)))+L(4)
1300 IF L(5)<0 THEN LET L(5)=L(5)+2*PI
1310 B(3)=ATN((S*TAN P(1)*SIN(L(5)-L(4)))/(S(3)*SIN(L(4)-L(3))))
1320 D(1)=ASN(SIN B(3)*COS O+COS B(3)*SIN O*SIN L(5))
1330 R(1)=ACS(COS B(3)*COS L(5)/COS D(1))
1340 IF -SIN O*SIN B(3)+COS O*COS B(3)*SIN L(5)>0 THEN LET R(1)=R(1)
1350 IF -SIN O*SIN B(3)+COS O*COS B(3)*SIN L(5)<0 THEN LET R(1)=2*PI-R(1)
1360 PRINT
1370 PRINT
1380 PRINT "DATA FOR HALLEY'S COMET:"
1390 PRINT
1400 PRINT "Right ascension:";TAB(20);DEG(R(1));" ";"degrees"
1410 PRINT "Declination:";TAB(20);DEG(D(1));" ";"degrees"
1420 H(1)=RAD(P(5)*15)-R(1)
1430 H(5)=H(1):D(5)=D(1)
1440 GOSUB 1760
1450 V(1)=V(5):U(1)=U(5):K(1)=V(1):Q(1)=U(1)
1460 PRINT
1470 PRINT "Azimuth:";TAB(20);DEG(V(1));" ";"degrees"
1480 PRINT "Altitude:";TAB(20);DEG(U(1));" ";"degrees"
1490 PRINT
1500 PRINT "Record values then press any key"
1510 PRINT "for a pictorial representation of the "
1520 PRINT "position of the comet..."
1530 KEYHIT=GET
1540 REM *COMET GRAPHICS*
1550 T(1)=ATN((Q(1)-Q(2))/(K(2)-K(1)))
1560 IF K(2)<K(1) AND Q(1)>Q(2) THEN LET T(1)=T(1)+PI
1570 IF K(2)<K(1) AND Q(2)>Q(1) THEN LET T(1)=T(1)+PI
1580 C=(80/R)*COS T(1):D=(80/R)*SIN T(1)
1590 CLS
1600 MOVE 10,10:DRAW 910,10
1610 MOVE 10,10:DRAW 10,910
1620 VDU 23,241,240,240,240,0,0,0,0,0
1630 L(5)=K(1)
1640 GOSUB 1710
1650 VDU 5
1660 LET K(4)=10+10*DEG(K(1)-(J-1)*PI/2)
1670 LET Q(4)=10+10*DEG(Q(1))
1680 GOSUB 1820
1690 VDU 4
1700 END
1710 IF L(5)>=0 AND L(5)<=PI/2 THEN LET J=1
1720 IF L(5)>=PI/2 AND L(5)<=PI THEN LET J=2
1730 IF L(5)>=PI AND L(5)<=3*PI/2 THEN LET J=3
1740 IF L(5)>=3*PI/2 AND L(5)<=2*PI THEN LET J=4
1750 RETURN
1760 IF H(5)>2*PI THEN LET H(5)=H(5)-2*PI
1770 IF H(5)<0 THEN LET H(5)=H(5)+2*PI
1780 U(5)=ASN(SIN D(5)*SIN L+COS D(5)*COS L*COS H(5))
1790 V(5)=ACS((SIN D(5)-SIN L*SIN U(5))/(COS L*COS U(5)))
1800 IF SIN H(5)>0 THEN LET V(5)=2*PI-V(5)
1810 RETURN
1820 MOVE K(4),Q(4)
1830 PRINT CHR$(241)
1840 MOVE 10,910:PRINT CHR$(57);CHR$(48)
1850 MOVE 40,40
1860 IF J=1 THEN PRINT CHR$(78)
1870 IF J=2 THEN PRINT CHR$(69)
1880 IF J=3 THEN PRINT CHR$(83)
1890 IF J=4 THEN PRINT CHR$(87)
1900 MOVE 910,40
1910 IF J=1 THEN PRINT CHR$(69)
1920 IF J=2 THEN PRINT CHR$(83)
1930 IF J=3 THEN PRINT CHR$(87)
1940 IF J=4 THEN PRINT CHR$(78)
1950 MOVE K(4),Q(4)
```

```
1960 DRAW K(4)-C,Q(4)+D
1970 RETURN
```

Spectrum

```
  10 REM "ALTAZ"
  20 PRINT
  30 PRINT
  40 PRINT "ALTITUDE AND AZIMUTH OF HALLEY's"
  50 PRINT "COMET AND THE SUN BETWEEN"
  60 PRINT "1st JAN. 1985 AND 31 DEC. 1986"
  70 DIM a(5): DIM b(5): DIM c(5): DIM d(5): DIM e(5): DIM f(5): DIM g(5)
  80 DIM h(5): DIM i(5): DIM k(5): DIM l(5): DIM m(12): DIM n(5)
  90 DIM p(5): DIM q(5): DIM r(5): DIM s(5): DIM t(5)
 100 DIM u(5): DIM v(5): DIM w(5): DIM x(5): DIM y(5): DIM z(5)
 110 LET RAD=PI/180
 120 LET DEG=180/PI
 130 REM *cumulative month days*
 140 FOR i = 1 TO 12
 150 READ m(i)
 160 NEXT i
 170 DATA 0,31,59,90,120,151,181,212,243,273,304,334
 180 REM *input constants comet*
 190 LET e(1)=0.967: LET w(1)=170.011: LET a(1)=17.94
 200 LET z(1)=58.1530: LET i(1)=162.238
 210 REM *input constants sun*
 220 LET w(2)=282.510396: LET g(2)=279.041470: LET o=23.44*RAD
 230 REM *input constants earth*
 240 LET e(3)=0.01672: LET w(3)=102.51044: LET g(3)=99.53431
 250 REM *input other constants*
 260 LET a=0.065709: LET c=1.002743
 270 PRINT
 280 INPUT "Observer's latitude(-90-+90 deg)",l
 290 IF l>90 OR l<-90 THEN GO TO 280
 300 LET l=RAD*l
 310 PRINT
 320 INPUT "Enter year (1985 or 1986)",y$
 330 IF y$<>"1985" AND y$<>"1986" THEN GO TO 320
 340 IF y$="1985" THEN LET b=17.359625
 350 IF y$="1986" THEN LET b=17.375540
 360 PRINT
 370 INPUT "Enter month (1=Jan...12=Dec)",n
 380 IF n<1 OR n>12 THEN GO TO 370
 390 PRINT
 400 INPUT "Enter day of month (1-31)",d
 410 IF d<1 OR d>31 THEN GO TO 400
 420 PRINT
 430 INPUT "Enter hour (0-23)",h
 440 IF h<0 OR h>23 THEN GO TO 430
 450 PRINT
 460 INPUT "Enter minute (0-59)",m
 470 IF m<0 OR m>59 THEN GO TO 460
 480 REM *FOR SUN*
 490 IF y$="1985" THEN LET x=3653+m(n)+d+(h+m/60)/24
 500 IF y$="1986" THEN LET x=4018+m(n)+d+(h+m/60)/24
 510 LET v=360*x/365.25
 520 IF v>=0 AND v<=360 THEN GO TO 560
 530 LET v=v-360
 540 IF v>=0 AND v<=360 THEN GO TO 560
 550 GO TO 530
 560 LET y=v+g(2)-w(2)
 570 IF y<0 THEN LET y=y+360
 580 LET y=RAD*y
 590 LET e=(360*e(3)*SIN y)/PI
 600 LET l(2)=v+e+g(2)
 610 IF l(2)>360 THEN LET l(2)=l(2)-360
 620 LET l(2)=RAD*l(2)
 630 LET d(2)=ASN (SIN o*SIN l(2))
 640 LET r(2)=ACS (COS l(2)/COS d(2))
 650 IF COS o*SIN l(2)>0 THEN LET r(2)=r(2)
 660 IF COS o*SIN l(2)<0 THEN LET r(2)=2*PI-r(2)
 670 PRINT
```

```
680 PRINT "DATA FOR THE SUN:"
690 PRINT
700 PRINT "Right Ascension:";TAB(16);DEG*r(2);" deg."
710 PRINT "Declination:";TAB(16);DEG*d(2);" deg."
720 LET x(5)=h+m/60
730 LET n(5)=m(n)+d
740 LET p(5)=n(5)*a-b+x(5)*c
750 IF p(5)>24 THEN LET p(5)=p(5)-24
760 IF p(5)<0 THEN LET p(5)=p(5)+24
770 LET h(2)=RAD*(p(5)*15)-r(2)
780 LET h(5)=h(2): LET d(5)=d(2)
790 GO SUB 1760
800 LET v(2)=v(5): LET u(2)=u(5): LET k(2)=v(2): LET q(2)=u(2)
810 PRINT
820 PRINT "Azimuth:";TAB(16);DEG*v(2);" deg."
830 PRINT "Altitude:";TAB(16);DEG*u(2);" deg."
840 PRINT
850 PRINT
860 IF DEG*u(2)>-18 THEN PRINT "Sky not dark"
870 REM *FOR COMET*
880 LET f=360*(x-4058.66)/(365.25*76)
890 IF f<0 THEN LET f=f+360
900 IF f>=0 AND f<=360 THEN GO TO 940
910 LET f=f-360
920 IF f>=0 AND f<=360 THEN GO TO 940
930 GO TO 910
940 LET f=RAD*f
950 LET q=f
960 LET f(1)=q-e(1)*SIN q
970 LET t=ABS (f-f(1))
980 IF t<0.0001 THEN GO TO 1020
990 LET c(1)=(f-f(1))/(1-e(1)*COS q)
1000 LET q=q+c(1)
1010 GO TO 960
1020 LET g=SQR ((1+e(1))/(1-e(1)))*TAN (q/2)
1030 LET n(1)=ATN g: LET n(1)=2*n(1)
1040 LET w(1)=RAD*w(1): LET z(1)=RAD*z(1): LET i(1)=RAD*i(1)
1050 LET l(1)=n(1)+w(1)
1060 LET r=(a(1)*(1-e(1)*e(1)))/(1+e(1)*COS n(1))
1070 LET g(1)=l(1)-z(1)
1080 LET p(1)=ASN (SIN g(1)*SIN i(1))
1090 LET l(4)=ATN (TAN g(1)*COS i(1))+z(1)
1100 IF g(1)>=PI/2 AND g(1)<=3*PI/2 THEN LET l(4)=l(4)+PI
1110 IF l(4)<0 THEN LET l(4)=l(4)+2*PI
1120 LET s=r*COS p(1)
1130 REM *FOR EARTH*
1140 LET v=(360*x)/(365.25*1.00004)
1150 IF v>=0 AND v<=360 THEN GO TO 1190
1160 LET v=v-360
1170 IF v>=0 AND v<=360 THEN GO TO 1190
1180 GO TO 1160
1190 LET y(3)=v+g(3)-w(3)
1200 LET y(3)=RAD*y(3)
1210 LET l(3)=v+(360*e(3)*SIN y(3))/PI+g(3)
1220 IF l(3)<0 THEN LET l(3)=l(3)+360
1230 IF l(3)>360 THEN LET l(3)=l(3)-360
1240 LET l(3)=RAD*l(3)
1250 LET w(3)=RAD*w(3)
1260 LET n(3)=l(3)-w(3)
1270 LET s(3)=(1-e(3)*e(3))/(1+e(3)*COS n(3))
1280 LET dif=l(3)-l(4)
1290 IF s<s(3) THEN LET l(5)=PI+l(3)+ATN ((s*SIN (dif))/(s(3)-s*COS (dif)))
1300 IF s>s(3) THEN LET l(5)=ATN ((s(3)*SIN (-dif))/(s-s(3)*COS (-dif)))+l(4)
1310 IF l(5)<0 THEN LET l(5)=l(5)+2*PI
1320 LET b(3)=ATN ((s*TAN p(1)*SIN (l(5)-l(4)))/(s(3)*SIN (l(4)-l(3))))
1330 LET d(1)=ASN (SIN b(3)*COS o+COS b(3)*SIN o*SIN l(5))
1340 LET r(1)=ACS (COS b(3)*COS l(5)/COS d(1))
1350 IF -SIN o*SIN b(3)+COS o*COS b(3)*SIN l(5)>0 THEN LET r(1)=r(1)
1360 IF -SIN o*SIN b(3)+COS o*COS b(3)*SIN l(5)<0 THEN LET r(1)=2*PI-r(1)
1370 PRINT
1380 PRINT
1390 PRINT "DATA FOR HALLEY'S COMET:"
1400 PRINT
1410 PRINT "Right ascension:";TAB(16);DEG*r(1);" deg."
1420 PRINT "Declination:";TAB(16);DEG*d(1);" deg."
1430 LET h(1)=RAD*(p(5)*15)-r(1)
1440 LET h(5)=h(1): LET d(5)=d(1)
```

```
1450 GO SUB 1760
1460 LET v(1)=v(5): LET u(1)=u(5): LET k(1)=v(1): LET q(1)=u(1)
1470 PRINT
1480 PRINT "Azimuth:";TAB(16);DEG*v(1);" deg."
1490 PRINT "Altitude:";TAB(16);DEG*u(1);" deg."
1500 PRINT
1510 PRINT "Record values then press CONT"
1520 PRINT "for a pictorial representation of the "
1530 PRINT "position of the comet..."
1540 STOP
1550 REM *comet graphics*
1560 LET t(1)=ATN ((q(1)-q(2))/(k(2)-k(1)))
1570 IF k(2)<k(1) AND q(1)>q(2) THEN LET t(1)=t(1)+PI
1580 IF k(2)<k(1) AND q(2)>q(1) THEN LET t(1)=t(1)+PI
1590 LET c=(10/r)*COS t(1): LET d=(10/r)*SIN t(1)
1600 CLS
1610 BORDER 0
1620 INK 7: PAPER 0
1630 PLOT 0,0: DRAW 175,0
1640 PLOT 0,0: DRAW 0,175
1650 LET l(5)=k(1)
1660 GO SUB 1710
1670 LET k(4)=1.8*DEG*(k(1)-(J-1)*PI/2)
1680 LET q(4)=1.8*DEG*q(1)
1690 GO SUB 1820
1700 GO TO 1950
1710 IF l(5)>=0 AND l(5)<=PI/2 THEN LET j=1
1720 IF l(5)>=PI/2 AND l(5)<=PI THEN LET j=2
1730 IF l(5)>=PI AND l(5)<=3*PI/2 THEN LET j=3
1740 IF l(5)>=3*PI/2 AND l(5)<=2*PI THEN LET j=4
1750 RETURN
1760 IF h(5)>2*PI THEN LET h(5)=h(5)-2*PI
1770 IF h(5)<0 THEN LET h(5)=h(5)+2*PI
1780 LET u(5)=ASN (SIN d(5)*SIN l+COS d(5)*COS l*COS h(5))
1790 LET v(5)=ACS ((SIN d(5)-SIN l*SIN u(5))/(COS l*COS u(5)))
1800 IF SIN h(5)>0 THEN LET v(5)=2*PI-v(5)
1810 RETURN
1820 PLOT k(4),q(4): PLOT k(4)+1,q(4): PLOT k(4),q(4)-1: PLOT k(4)+1,q(4)-1
1830 PRINT AT 0,0;CHR$ 57;CHR$ 48
1840 IF j=1 THEN PRINT AT 21,0 CHR$ 78
1850 IF j=2 THEN PRINT AT 21,0 CHR$ 69
1860 IF j=3 THEN PRINT AT 21,0 CHR$ 83
1870 IF j=4 THEN PRINT AT 21,0 CHR$ 87
1880 IF j=1 THEN PRINT AT 21,21 CHR$ 69
1890 IF j=2 THEN PRINT AT 21,21 CHR$ 83
1900 IF j=3 THEN PRINT AT 21,21 CHR$ 87
1910 IF j=4 THEN PRINT AT 21,21 CHR$ 78
1920 PLOT k(4),q(4)
1930 DRAW -c,d
1940 RETURN
1950 REM END OF PROGRAM
```

12   How close will it come to the Earth?

At its last appearance in 1910 Halley's comet came very close to Earth and our planet actually passed through the tail of the comet for a few hours. There is no direct evidence that anything out of the ordinary happened to the general population except for the alarm generated prior to the event (section 27). There are, however, reports that the night sky was abnormally bright and that the astronomers at the Pic du Midi Observatory, high in the French Pyrénées, noted haloes round the Sun and round the Moon. The latter could have been the result of a nebulous cloud of minute particles in the upper atmosphere such as might be present in the comet's tail. The nearest approach of Halley's comet to the Earth was only 22 million km or 0.15 a.u.

It so happened on this occasion that Comet Halley passed directly between the Sun and the Earth two days earlier. Theoretically it should have been visible to astronomers with the correct equipment as it made its transit across the Sun's disc, but it could not be seen, which indicates the tenuous nature of the comet.

In fact Halley's comet has made as many as six approaches to the Earth of less than 0.1 a.u. since 918 BC, the nearest approach being only 0.04 a.u. in AD 837. This near approach compares almost exactly with the comet IRAS-Iraki-Alcock of May 1983 which was seen with the naked eye by many non-astronomers and studied by radar with the radio telescope at Arecibo (see section 29).

This time Halley's comet will be at its closest to the Earth (0.42 a.u.) on 12 April 1986 after the comet's passage round the far side of the Sun. Its nearest approach to the Sun, its perihelion passage, will be larger than this (0.59 a.u.) and will occur on 9 February 1986.

Apart from very minor deviations, Halley's comet moves in a path which could be drawn on a flat plane. The same is true of the Earth's path round the Sun. The two planes are inclined at an angle of only 18 degrees, so the representation of the motions of the Earth and comet shown in the diagram in figure 23 is not too distorted.

Note that Halley's comet moves in a very elongated ellipse (eccentricity 0.967) (see figure 5), extending beyond the orbit of Neptune. The geometry of the orbit is such that when the comet is near the Sun its orbit may be treated as parabolic. Advantage has been taken of the approximation in the computer program given in section 25. It demonstrates the relative positions of Halley's comet and the Earth at intervals of one month.

If you do not have access to a microcomputer, or indeed if you do, you might like to see the same information by making the model described in section 26. This has the advantage that the orbits can be shown at their correct relative orientations in space.

An important point is that Comet Halley moves round the Sun in the direction opposite to that of the Earth in *its* motion round the Sun. Its motion is *retrograde*. This has crucial significance for the Giotto space probe (section 30).

13   How best can it be observed?

The factors which will determine whether or not you will see Halley's comet are numerous.

Comet Halley, like other comets, will be brightest when it is nearest the Sun. This in itself inhibits observation at what would appear to be the best time, because the Sun's strong rays mask the weaker ones of the comet. Nevertheless it *is* possible to take advantage of this increased brightness near the Sun; for if the comet is to the east of the Sun it might be visible in the evening twilight soon after the Sun has set, when the Sun's rays themselves are masked by the Earth. Similarly, if the comet is to the west of the Sun it might be visible in the morning twilight just before the Sun rises. Technically, *astronomical twilight* begins in the evening when the Sun has dipped 18° below the horizon. This is about 70 minutes after the setting of the Sun, but it varies according to one's latitude. There is the additional factor that the atmosphere near the horizon is often not clear.

Although Halley's comet will be less bright when it appears to be further from the Sun, it will be located in a darker sky for some time after the Sun has set, or before the Sun has risen, as the case may be.

The unaided eye has only a limited light-gathering capability determined by the size of the pupil. It is however remarkably sensitive, and there is a good chance that you will see Halley's comet if the magnitude is not greater than 5. Remember, the greater the magnitude the fainter the object. Your eyes adapt themselves to the dark, achieving a much more sensitive state in about ten minutes by covering their retinae with a purple substance, which absorbs light and is at once decomposed by it. This chemical decomposition is detected by the nerves and the appropriate signals are sent to the brain. So even a brief return to a lighted room or the injudicious use of a torch reduces your likelihood of seeing Halley's comet, especially if it is faint. Once you have located it, you may see it better by looking slightly to one side.

If we increase our light-gathering capacity by using binoculars, whose objective lenses are much larger than the pupils of our eyes, we increase our ability to see Halley's comet. At the same time, it is a fallacy to think that by increasing the magnifying power of our binoculars we necessarily increase the visibility of the comet. Only a specific amount of light comes from the comet, and when we magnify this unduly the light is effectively spread over a bigger area and the comet may seem fainter. For the observation of comets of magnitude 4 to 8 some authorities recommend 10 × 50 or 20 × 80 binoculars, the first figure in each case being the magnification and the second the diameter of the ojective lens in millimetres.

If you are an amateur astronomer you will already know all this, and may have experienced it. If you are not, you will be limited to using whatever equipment you have, including your unaided eyes, but the International Halley Watch Committee broadsheet indicates that 7 × 35 or 7 × 50 binoculars would be useful.

The remarks on binoculars apply also to the use of telescopes. It is likely that your telescope, if you have one, will have even greater light-gathering power because of the large objective lens or primary mirror. If Halley's comet has a magnitude of 8 to 10, a telescope with an aperture of 15–20 cm (6–8 in.) should be very satisfactory. For magnitudes of 10 to 12 use apertures of 25–30 cm (10–12 in.) though one authority considers that 8 in. is large enough for magnitude 12. For magnitudes of 12 to 14 use apertures of 35–50 cm (14–20 in.). These large sizes for amateurs indicate the use of a reflecting telescope (see figure 27) with a large mirror. The magnifying power of a given telescope is altered by changing the eyepiece. Low-power eyepieces, say those which give a magnification of 2 for every centimetre of diameter of the objective lens or primary mirror, will be found to be most satisfactory.

The use of low magnification will allow you to see more of the sky around Halley's comet, which is useful if you are searching. The use of high magnification, especially with binoculars, makes the sighting more difficult, not only because any slight movement of the binoculars causes the field of view to move unduly, but also because of the small field of view.

Three other factors may affect the ease with which you may see the comet. If you are in a locality with artificial street lighting you will need to get away from it. The amount of cloud present may have the effect of reducing the apparent brightness of the comet, as also may be the presence of the Moon, particularly the full Moon. It will be full on 29 August, 28 September, 27 October, 26 November, 25 December 1985 and on 24 January, 22 February, 24 March, 22 April, 22 May, 20 June, 20 July, 18 August, 17 September 1986. For a computer program to determine the date of full Moon or new Moon, or indeed any phase of the Moon, in any specified month of any year, see section 15. The program will give this information to an accuracy of one day, though a very few dates may be two days in error. For this return of Halley's comet the calendar will be Gregorian (G at the appropriate prompt in all the programs).

Our calendar is based on the position of the Sun against the background of stars. About 21 March in the spring of each year the number of hours of daylight and the number of hours of darkness are equal, so we refer to this date as the *vernal equinox*. We observe the Sun in the same position one *tropical* year later. On this definition and with our specified length of a day the tropical year comes out at rather more than 365 days (365.242). Consequently the seasons slowly get out of step with the dates. To compensate for this, Julius Caesar in 46 BC introduced the *leap year* so that an extra day was included in every fourth year. On average the length of the calendar year (365.250) is about 11 minutes longer than the tropical year. By the sixteenth century the two were out of step by a total of about 10 days. As a result Pope Gregory XIII decreed that 4 October 1582 AD should be followed by 15 October 1582. (This caused rioting because people really believed that by this device their lives had been shortened.) The calendar year was then made to accord almost exactly with the tropical year by keeping the leap year when the year is divisible by 4, but years ending in 00 (which are always divisible by 4) are only counted as leap years if the year is divisible by 400. This Gregorian calendar is still in use today. For dates after 1582 we work on the Gregorian calendar and it is customary to give dates before 1582 as Julian dates, even those before 46 BC.

You might like to check the Chinese record which states that a comet, which undoubtedly was Halley's comet, appeared on 21 September 1607 pointing NW. The record says the Moon was new. The program gives 20 September for the new Moon in September 1607 AD. For full explanatory notes on all the computer programs see appendix A.

14   Observing conditions for Halley's comet

It is possible to calculate the expected altitude, the angle above the horizon, of Halley's comet at a particular place on any specified date both at the time when the Sun is setting in the evening and when the Sun is rising in the morning. If the altitude of Comet Halley is high when the Sun sets there will be a long period to observe the comet in a dark sky before the comet itself sets. If the comet's altitude is high when the Sun rises there will have been a long period of observing in the dark sky before sunrise.

(1) We may consider that on any particular date the most favourable latitude on Earth from which to observe Halley's comet is the latitude which gives us the longest observing time in a dark sky. Assuming this definition of the most favourable observing latitude, we have used the method described by Dr J. B. Tatum of the University of Victoria, British Columbia, Canada, to produce the graphs shown in figures 10 and 11. From them we can read the most favourable latitude from which to observe Halley's comet for any date on which the comet will be at naked-eye brightness. Thus, from figure 10, the most favourable latitude, as defined above, on 29 January 1986, is about 40°N, the latitude of Rome, Madrid, Lisbon, New York and Tokyo. Also from figure 10, on 25 March 1986 we obtain the most favourable latitude as about 30°S, the latitude of Cape Town and Sydney.

Again, it will be clear from figure 10 that Halley's comet may best be observed at places in the northern hemisphere *until* perihelion passage of the comet and at places in the southern hemisphere *after* perihelion passage.

We may refer to the graphs in figure 11 to find what the altitude of Halley's comet will be at sunset or sunrise on a particular date at places with the most favourable latitude.

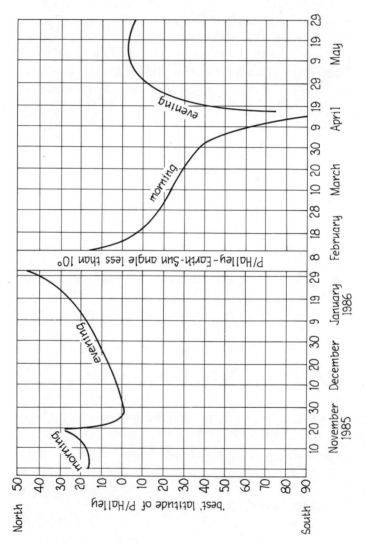

FIGURE 10 The 'best' latitude at which to see Halley's comet.

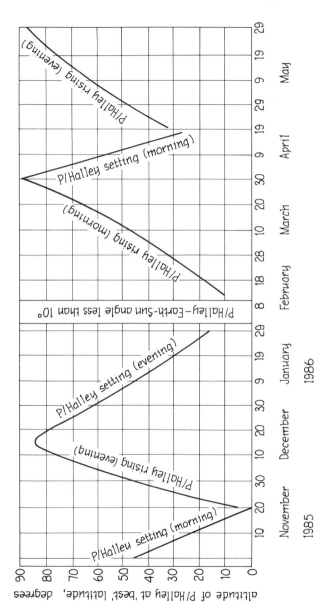

FIGURE 11   The altitude of Halley's comet at the 'best' latitude.

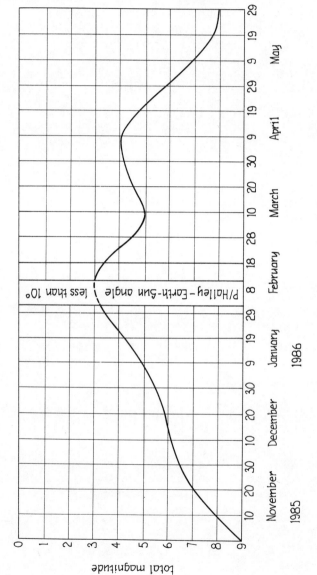

FIGURE 12   *The magnitude of Halley's comet.*

Thus, from figure 11 on 29 January 1986 we find that the altitude of Comet Halley will be 15° at sunset at places with a latitude of about 40°N. Also from figure 11 on 25 March 1986 Halley's comet will have achieved an altitude of 77° in the dark sky before sunrise at places with a latitude of about 30°S.

From figure 12 we can see that the magnitude of Halley's comet on 29 January should be about 3.5 and on 25 March 1986 about 4.5, both well within naked-eye visibility.

(2) For most of us the discussion in (1) above will be academic, but the general conclusions are of interest. Nevertheless professional astronomers do travel to observatories throughout the world and some will have the points in (1) in mind. Also the British Interplanetary Society is organizing a trip to South Africa, latitude about 30°S, during the first part of April 1986 to observe Halley's comet. Figure 10 indicates that this is a most favourable latitude at that time. The Executive Secretary of the BIS would be pleased to give details of the trip (telephone 01-735-3160).

(3) The sky is not dark as soon as the Sun sets, nor is it dark until the Sun rises because the Earth's atmosphere bends the Sun's rays and gives us the twilight period. As mentioned at the beginning of this section twilight ends when the Sun has set about 18° below the horizon in the evening and begins when the rising Sun is still about 18° below the horizon. One of the figures 13–18, the one corresponding most nearly to our own latitude, can be used to indicate for how many hours Halley's comet will be visible in a dark sky on any date when the comet should be at naked-eye brightness.

To extract this information on a particular date we must follow the vertical line through that date at the bottom. Note that as we pass zero (or 24) hours on the time axis we

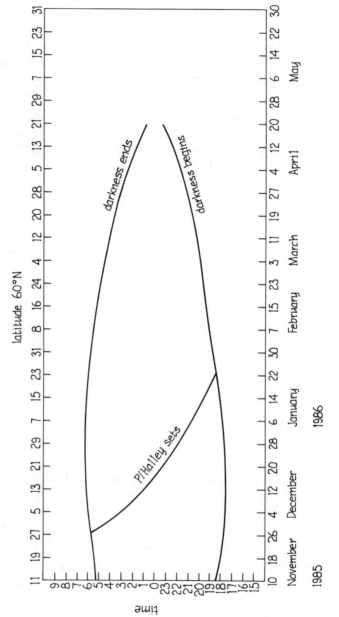

latitude 60°N

FIGURE 13 *The visibility of Halley's comet at latitude 60°N.*

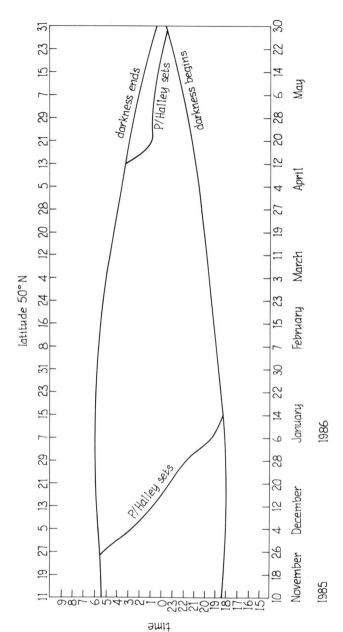

latitude 50° N

FIGURE 14 The visibility of Halley's comet at latitude 50° N.

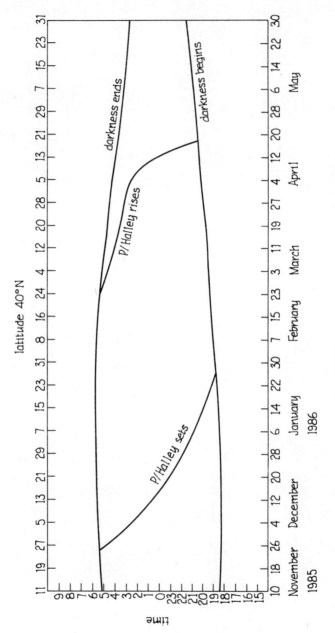

latitude 40°N

FIGURE 15   The visibility of Halley's comet at latitude 40°N.

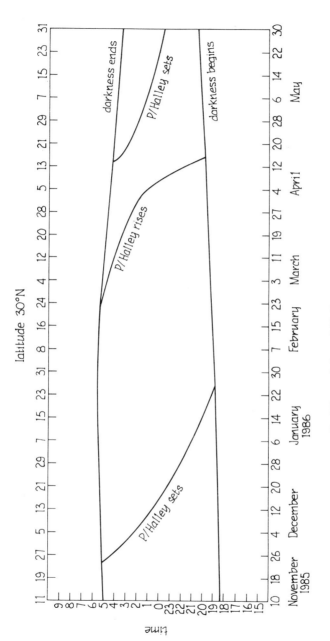

latitude 30°N

FIGURE 16   The visibility of Halley's comet at latitude 30°N.

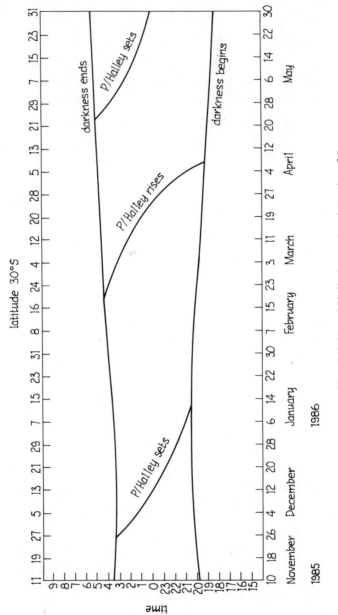

latitude 30°S

FIGURE 17   *The visibility of Halley's comet at latitude 30°S.*

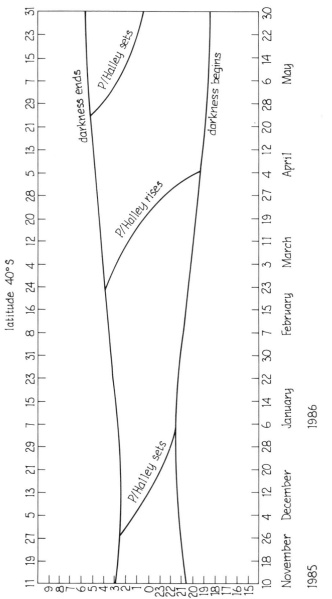

FIGURE 18 The visibility of Halley's comet at latitude 40°S.

are passing into the next day whose date is given at the top. In general, the number of hours between the 'darkness begins' line and a 'P/Halley sets' line indicates the length of time Halley's comet is above the horizon in a dark sky. Thus on 20 December 1985 in latitude 40°N (figure 15) darkness falls at 1820 hours and P/Halley sets 4.7 hours later at 2300 hours. Similarly, the number of hours between a 'P/Halley rises' line and the 'darkness ends' line indicates the length of time Halley's comet is above the horizon in a dark sky. Thus the vertical progression from 8 April 1986 at the bottom indicates that P/Halley rises at 0150 hours on 9 April 1986 and darkness ends at 0400 hours on the same day, giving 2.2 hours visibility in a dark sky. Note, however, that in latitude 30°S Halley's comet should be visible in a dark sky for the whole night from 8 April to 22 April 1986. This is especially relevant for the BIS trip to South Africa mentioned above.

15   Determining the dates of any phase of the Moon: computer programs

BBC Model B

```
10 REM "MOON"
20 CLS
30 PRINT "MOON PHASE"
40 DIM R(19)
50 DIM L(33)
60 DIM M(12)
70 FOR I = 1 TO 19
80 READ R(I)
90 DATA 9,25,10,26,12,28,13,29,15,1,17,3,19,4,20,6,22,7,23
100 NEXT I
110 FOR I = 1 TO 33
120 READ L(I)
130 DATA 9,28,17,6,25,14,3,22,11,30,19,8,27,16,5,24,13,2,21,10,29
140 DATA 18,7,26,15,4,23,12,1,20,9,28,17
150 NEXT I
160 FOR I = 1 TO 12
170 READ M(I)
180 DATA 0,2,2,4,4,6,7,8,9,10,11,13
190 NEXT I
200 PRINT
210 REPEAT INPUT "Year ",Y: UNTIL Y>=0
220 LET YR=Y
230 REPEAT INPUT "BC or AD",A$: UNTIL A$="AD" OR A$="BC"
240 REPEAT
250 PRINT "Input month number, enter 11 for JAN,"
260 PRINT "12 for FEB, 1 for MARCH...10 for DEC:"
```

```
270 INPUT Z
280 UNTIL Z>=1 AND Z<=12
290 REPEAT INPUT "Input J (Julian) or G (Gregorian) ",B$
300 UNTIL B$="J" OR B$="G"
310 REPEAT INPUT "Enter phase: 0 for new, 15 for full",C
320 UNTIL C>=0 AND C<=29
330 IF Z>10 THEN LET Y =Y-1
340 IF A$="AD" THEN LET J =Y+4712
350 IF A$="BC" THEN LET J=4713-Y
360 LET Q=INT (J/76)
370 LET N=J-Q*76
380 LET S=INT (Q/4)
390 LET R= INT (N/4)
400 LET L =N-R*4
410 FOR I = 1 TO 30
420 IF L(I)=R(R+1) THEN LET D=L(I+L)
430 NEXT I
440 LET W=D-S-M(Z)+C
450 IF B$="J" THEN LET T=W
460 IF Y>1582 AND Y<=1700 THEN LET Y=10
470 IF Y>1700 AND Y<=1800 THEN LET Y=11
480 IF Y>1800 AND Y<=1900 THEN LET Y=12
490 IF Y>1900 THEN LET Y=13
500 IF B$="G" THEN LET T=W+Y
510 IF T<0 THEN LET T=T+30
520 IF T>30 THEN LET T=T-30
530 PRINT
540 PRINT "A moon ";C;" days old occurs on"
550 PRINT "Day ";T;", month ";Z;", year ";YR;" ";A$
560 GOTO 200
```

Spectrum

```
 10 REM "MOON"
 20 CLS
 30 PRINT "MOON PHASE"
 40 DIM r(19)
 50 DIM l(33)
 60 DIM m(12)
 70 FOR i = 1 TO 19
 80 READ r(i)
 90 DATA 9,25,10,26,12,28,13,29,15,1,17,3,19,4,20,6,22,7,23
100 NEXT i
110 FOR i = 1 TO 33
120 READ l(i)
130 DATA 9,28,17,6,25,14,3,22,11,30,19,8,27,16,5,24,13,2,21,10,29
140 DATA 18,7,26,15,4,23,12,1,20,9,28,17
150 NEXT i
160 FOR i = 1 TO 12
170 READ m(i)
180 DATA 0,2,2,4,4,6,7,8,9,10,11,13
190 NEXT i
200 PRINT
210 INPUT "Year ",y
220 IF y<0 THEN GO TO 210
230 LET yr=y
240 INPUT "BC or AD",a$
250 IF a$="ad" THEN LET a$="AD"
260 IF a$="bc" THEN LET a$="BC"
270 IF a$<>"BC" AND a$<>"AD" THEN GO TO 240
280 INPUT "Enter month: (11=Jan, 12=Feb,  1=Mar...10=Dec",z
290 IF z<1 OR z>12 THEN GO TO 280
300 INPUT "Enter J(ulian) or G(regorian)",b$
310 IF b$="j" THEN LET b$="J"
320 IF b$="g" THEN LET b$="G"
330 IF b$<>"J" AND b$<>"G" THEN GO TO 300
340 INPUT "Enter phase: 0(new) to 15(full)",c
350 IF c<0 OR c>29 THEN GO TO 340
360 IF z>10 THEN LET y =y-1
370 IF a$="AD" THEN LET j =y+4712
380 IF a$="BC" THEN LET j=4713-y
390 LET q=INT (j/76)
```

```
400 LET n=j-q*76
410 LET s=INT (q/4)
420 LET r= INT (n/4)
430 LET l =n-r*4
440 FOR i = 1 TO 30
450 IF l(i)=r(r+1) THEN LET d=l(i+1)
460 NEXT i
470 LET w=d-s-m(z)+c
480 IF b$="J" THEN LET t=w
490 IF y>1582 AND y<=1700 THEN LET y=10
500 IF y>1700 AND y<=1800 THEN LET y=11
510 IF y>1800 AND y<=1900 THEN LET y=12
520 IF y>1900 THEN LET y=13
530 IF b$="G" THEN LET t=w+y
540 IF t<0 THEN LET t=t+30
550 IF t>30 THEN LET t=t-30
560 PRINT
570 PRINT "A moon ";c;" days old occurs on"
580 PRINT "Day ";t;", month ";z;", year ";yr;" ";a$
590 GO TO 200
```

16　How do we know?

Historical evidence, in the form of literature, photographs and drawings, that Halley's comet has been returning to the neighbourhood of the Sun and Earth at intervals of about 76 years since 239 BC is abundant. Dr Joseph Brady, Dr T. Kiang and others have extrapolated backwards in time to determine possible dates of appearances of Halley's comet for almost 1000 years before that. Their calculations have allowed them to associate references in the literature more readily with Halley's comet, for there are innumerable references to other comets too.

The further we go back in time, however, the more uncertain the correlation, partly because the descriptions of comets vary so much and partly because the various calendars – Gregorian, Julian, Babylonian, Egyptian, Greek, Chinese and Japanese – have been confused, not only among one another but also within themselves.

Thus we have *aster kometes*, the Greek origin of our our word 'comet', but really meaning 'long-haired star'. Halley himself referred to *astrum barbatum*, the 'bearded star' in his 'Ode to Newton' in the *Principia*. The Chinese records used 'brush stars', 'sailing stars', 'candle-stars' to name but a few, but the favoured technical term was *hui*, 'brush stars'.

The Chinese *po* referred to a comet seen from the Earth in the direction of the Sun so that the comet appeared to have no tail. Reference to the listing below will reveal other graphic descriptions.

We are largely dependent on Chinese observations of Halley's comet before AD 1400 for positive identification. Most European records are not sufficiently accurate before that date. The drawings and paintings of Halley's comet (see plate 13) vary considerably with the impression made on the artist, though it is generally agreed that Giotto's painting (plate 6) gives one of the most accurate representations among the earlier pictures.

From about AD 1840 photography was available to astronomers, so we have accurate records of the 1910 appearance (plates 7 and 8). Charles Piazzi Smyth made some excellent drawings (plate 3) at the Cape of Good Hope of Halley's comet in 1835 just before the invention of the *daguerreotype*. In general, Halley's comet has favoured the southern hemisphere for observation. A photograph of the present reappearance was secured in October 1982 (plate 12) though it may be difficult to recognize the impression as Halley's comet because the techniques used to secure it are quite different from the ones to which we are accustomed.

To describe each known apparition in detail would be too extensive, but we give below a description of some of the more interesting features of some previous sightings.

1986
Detected at Palomar Observatory on 16 October 1982 by David C. Jewitt and G. E. Danielson at a distance of 1635 million km from the Earth.

1910
Dr Max Wolf found an image of Halley's comet on a photographic plate taken at Konigstuhl Observatory on

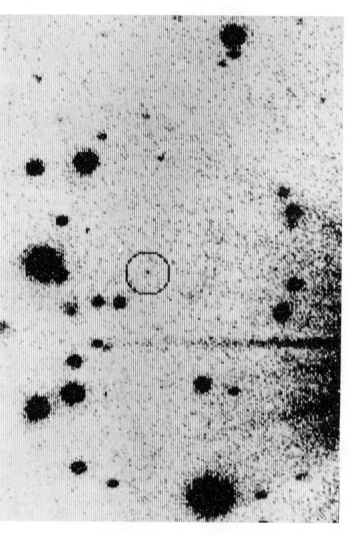

PLATE 12 Halley's comet recovered on 16 October 1982 with the 200-inch telescope at Mount Palomar. The image of the comet is at the centre of the circle.

12 September 1909. Comet Halley was then at a distance of 480 million km.

1835

First observed on 6 August 1835 by Dumouchel, in Rome. Sir John Herschel left us some fine drawings of this very bright appearance of Halley's comet.

1759

This is the reappearance which Halley forecast, for the first time in history. Halley died in AD 1742 and did not see it, but it was observed all over the world by many astronomers including Charles Messier, whose list of special objects in the sky is still famous – the Andromeda nebula or galaxy is M31 on this list. Comet Halley was picked up on 25 December 1758 by a German amateur astronomer Johann Palitzch.

1682

Halley himself observed this appearance. His famous prediction of its return in 1758 is described in section 4 of this book.

1607

Johannes Kepler and Christian Longomontanus, assistant of Tycho Brahe, observed this apparition. All the later observations of Halley's comet could be made with a telescope, probably only just invented about this time. Kepler's pre-telescopic observations are of such high accuracy as to have been referred to by Bessel as 'little gold nuggets'.

1531

Observed and meticulously recorded by Peter Apianus of Ingolstadt in his *Astronomicum Caesareum* (plate 5). Apianus observed that the tail of a comet always points away from the Sun along the comet–Sun line (in spite of showing the Sun below the horizon in nearly all his diagrams!). In this he was in agreement with Girolamo Frascatoro, a contemporary Italian who observed the same thing. Apianus' observations were regarded by Halley as uncertain.

1456
The fear of comets was emphasized at this bright return. The Pope, Calixtus III, publicly excommunicated it as an agent of the Devil. Paolo Toscanelli recorded 24 observed positions, and in this case the identification is sound. The Turks were at this time threatening to overrun Europe, and the appearance of Halley's comet was associated by some with the fall of Constantinople (but see section 6).

1378
This appearance was not very brilliant, but the Chinese record that it was seen between 10 September and 10 November.

1301
This spectacular apparition of Halley's comet was commented on by many contemporary historians. Giovanni Villani wrote, in *Chroniche Storiche*, of a comet 'with great tails of fumes behind it'. The painter Giotto di Bondone included an accurate representation of it in a detail of his painting *The Adoration of the Magi* (plate 6).

1222
There is no doubt that this return was observed in Europe. The Chinese say that its structure looked like that of Jupiter.

1145
Halley's comet made a 'poor showing' on this occasion. Nevertheless Eadwine, a monk at Canterbury, while making a copy of Psalm 5 from the ninth-century Utrecht Psalter was sufficiently diverted from his task by it. He made a crude drawing of the comet at the bottom of the page (plate 13). He also added a Saxon text about the

comet. (A person who writes in the margins of books is known as a 'glosser'.)

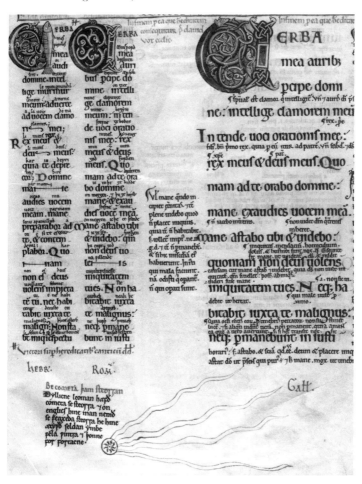

PLATE 13　*Psalm 5 copied by the monk Eadwine from the Utrecht Psalter. The comet at the bottom of the page is Halley's comet 1145. The Anglo-Saxon text says that in English it is called the long-haired star, and it appears seldom ('after many winters'). It is seen as a portent. (Translation kindly provided by Dr Bruce Mitchell.)*

1066
Halley's comet appeared at the time of the Battle of Hastings and is illustrated in the Bayeux Tapestry (plate 10). It featured a long tail.

989
This bluish-white apparition is recorded in the Saxon Chronicles.

912
Dr Kiang suggests that because of the uncertainty of dates the Chinese could have seen a fragment of Halley's comet travelling three months ahead of the main body.

837
This must have been one of the most spectacular returns observed. Halley's comet made its closest approach top the Earth. It moved over 60 degrees westward on 9/10 April and its tail spread well over 90 degrees.

AD *760, 684, 607, 530, 451, 374, 295, 218, 141 and 66*
All these apparitions can be found in the Chinese records. The translations give the lengths of the tail in feet! It would seem that 10 ft (1 *chang*) means a spread of about 12 degrees. The appearance of AD 684 is noteworthy because the earliest known drawing of Halley's comet, printed from a woodcut in the *Nuremberg Chronicles* of Hartmann Schedel (1483), refers to that apparition. It is clear that it is not a faithful representation. Flavius Josephus in his 'Jewish War' refers to the AD 66 apparition as a 'broadsword-shaped star over Jerusalem', which remained for a whole year.

11, 86, 163 and 239 BC
For these returns of Halley's comet we must again refer to Chinese records, except that there appears to be no record of the 163 BC return. The year 239 BC is probably the

earliest return found in any records: in that year Halley's comet was first seen in the east, then in the north and then for 16 days in the west.

17 Photographing Halley's comet

If Halley's comet comes up to expectations and you are able to locate it easily in a dark sky you might like to try to photograph it. No special equipment is needed other than your camera and a rigid tripod on which to mount it. If you have a cable release so much the better. This will help to keep camera vibration down to a minimum when you open and close the shutter.

The International Halley Watch leaflet on *Seeing Halley's Comet* gives brief advice for an attempt at photography. A camera of focal length between 28 mm and 200 mm would be suitable. Use the lowest *f* number setting possible and to keep the shutter open adjust the time setting to B. The distance setting will of course be infinity. The film should be a fast black and white or colour film. Reasonable exposure times should be in the range 10 seconds to 10 minutes.

The advice given by Douglas Arnold in the British Interplanetary Society publication *Spaceflight* is more positive. He says that 400 ASA colour or black and white film will be a must. If the comet is bright you can start an exposure programme from about 1/60 second up to about 2 minutes with a moderate aperture of around $f2.8$, but conditions may dictate a longer exposure from 2 minutes to perhaps 20 minutes.

When your eyes become adapted to the dark they have a low sensitivity to colour, but colour film does not have this deficiency. Kodak recommend you to try Kodacolor 400 for colour prints or Kodak Ektachrome 200 (Daylight) for colour slides.

The sky appears to rotate from east to west through south at the rate of 15 degrees per hour. To the naked eye this goes on imperceptibly, but it is important when looking through a telescope, and more important when attempting photography. If your camera is stationary, simple geometry will tell you that a 2-minute exposure with a camera of 100 mm focal length will result in a movement of almost 1 mm on the film for typical distances of the comet. Since this movement is continuous the result will be a blurred image of the comet head. The same will apply to stars in the field of view, which will appear as line images on the film instead of points of light.

A guided camera is therefore desirable. If you can fix the camera piggy-back on to a telescope you can adjust the telescope smoothly to keep the image of the comet in the centre of the field of view throughout the exposure. Alternatively, you can make the very simple but very effective adjustable camera mount described below.

For shorter exposures you may obtain reasonable images without guiding, provided your camera is rigidly mounted and you use your cable release. For the longer exposures, even if you guide on the comet and gain a sharp photograph of it, the stars may still appear as short bright lines since the comet is moving slowly relative to the stars, in addition to having the daily motion common to both.

It is suggested by Douglas Arnold that you will require (a) plenty of film, (b) plenty of patience and (c) guiding facilities for the longer exposures. You can, of course, adopt the good old engineers' principle – try it and see! As Barnes Wallis said in a much more serious context, 'why not?'

A simple guiding device for photographing Halley's comet

The main requirement for the longer time exposures desirable for photographing Halley's comet is the need to move the camera steadily in such a way as to counteract the effects of the Earth's daily motion. The Earth revolves on an axis very nearly parallel to the line from the observer to the Pole Star. If we can mount a camera so that we can rotate it at the appropriate speed about an axis parallel to this, objects in the sky will remain stationary on the film. This can be achieved by the easily-made device shown in figure 19 and plate 14.

The construction of the device is shown in figure 19 and plate 14, except for a few important details. Each of the boards is about 300 mm × 150 mm × 12 mm thick. The screw thread of the bolt holding the camera should of course be compatible with that in the camera base to avoid damage to the camera. The guiding bolt screws through the back board in a slightly undersized hole 290 mm from the hinge line. (Alternatively, a nut may be fixed in the back board in this position.) A ¼-in. diameter Whitworth bolt

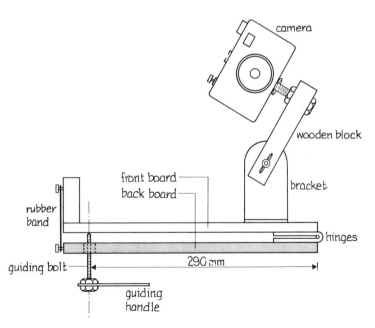

FIGURE 19   *Guiding device for photographing the comet.*

PLATE 14   *A guiding device for photographing Halley's comet.*

driven at 1 revolution per minute will push the front board forwards at the correct rate. The rubber band restrains the front board movements.

The back board should be fixed to a heavy anchor block for stability. The block should be cut at an angle to the horizontal equal to the observer's latitude and orientated so that the hinge line points to the Pole Star. The two wing nuts then allow the camera to be pointed at Halley's comet. First practise turning the guiding handle at the correct rate by taking few photographs of a recognizable constellation of stars. George Haig has taken an excellent colour transparency of the constellation Orion using this apparatus and an exposure of 5 minutes at *f*2.8 on High Speed Ektachrome film.

III

The Science of Comets

18   What *is* a comet?

This question is put in general form because our knowledge of Halley's comet is derived not only from our inspection of it but largely from our study of other comets too.

The spectrum (plate 15) of a comet formed by light normally visible to the human eye is relatively easy to obtain and record photographically using a telescope (see section 23). The spectrum extends, as a continuous band of the colours of the rainbow, from red at one end to violet at the other. This *continuous spectrum* is crossed by dark *absorption lines* and is formed from sunlight reflected from the comet. Due to the changes taking place in the atoms and molecules in the comet, explained in greater detail in section 19, the continuous spectrum is also crossed by a number of bright *emission lines*, the positions of which are characteristic of the atoms or molecules causing them.

Radiation exists outside the range visible to the naked eye and contributes to an extended spectrum. Some of the radiation which extends the violet end of the spectrum is known as *ultra-violet* radiation. The red end of the spectrum is extended into the *infra-red* region, while further out is the *radio* region.

Improved instruments and observational techniques

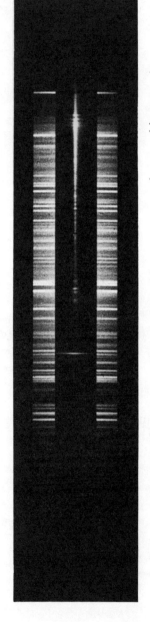

PLATE 15   *Spectrum of Halley's comet (centre). The calibration spectra at the top and bottom assist in the identification of elements, compounds and ions in P/Halley's spectrum which clearly shows emission lines.*

using these extended regions of the spectrum have led to the detection and identification of more atoms and molecules in comets.

Ultra-violet radiation is largely filtered out by the Earth's atmosphere. By placing ultra-violet sensitive equipment on board the Orbiting Astronomical Observatory artificial earth satellite, above the Earth's atmosphere, it was discovered that some comets have a head which is surrounded by a cloud of atomic hydrogen of radius about one million km. At the other end of the spectrum, at infrared wavelengths, thermal radiation from the heated dust in the tail has been detected. Observations using radio telescopes detected the presence of two organic molecules, methyl cyanide (CH_3CN) and hydrogen cyanide (HCN), in that visually disappointing comet Kohoutek.

To sum up, the following atoms and molecules have been detected in the head of comets:

$$H, C, C_2, C_3, CH, CN, {}^{12}C, {}^{13}C, HCN, CH_3CN, NH,$$
$$NH_2, O, OH, H_2O, CO, CS, S, N, Si, Al, Y, K, Na,$$
$$Ca, Cr, Co, Mn, Fe, Ni, Cu, V.$$

Single capital letters (e.g. H, Si) refer to atoms; combinations (e.g. HCN, CO) refer to molecules, while ${}^{12}C$ and ${}^{13}C$ refer to isotopes of carbon. Quite an impressive array! The last nine are metals which can be observed in the spectra of bright comets only. The sodium (Na) emission is sometimes so strong that it gives a yellow appearance to the comet. It was the discovery of the cyanogen (CN) which gave rise to the alarm at the last appearance of Halley's comet (section 27).

The tails of comets give positive ions (atoms or molecules with one electron removed; see section 19) only:

$$CH^+, CO^+, CO_2{}^+, N_2{}^+, OH^+, C^+, DH^+, Ca^+,$$

in addition to spectra indicating the presence of silicate dust particles.

How then is all this material held together in such a way that it can not only survive a close approach to the Sun but also survive a round trip of 76 years reaching out into the very cold regions of the outer solar system, and do this repeatedly?

For a time the favourite theory, that of Fred Whipple, was that the material was held together by water-ice. This so-called 'dirty-snowball' model allows for the evaporation of the outer layers of a nucleus as the comet approaches the Sun, releasing dust and the gases which form the components of the tail. These are driven in a direction away from the Sun by the pressure exerted by light from the Sun and also by the solar wind (see section 7).

This description generally fits observation, but Professor Raymond Lyttleton has raised numerous objections to some of the details. No one can have observed the nucleus because from mass considerations it cannot be much larger than 1 km in radius; and this is expected to produce not only a large coma and tail but also the extensive hydrogen envelope mentioned above. Further objections can be raised in that in many comets the coma has been observed to *contract* as the comets approached the Sun and to expand again when they recede.

Professor Lyttleton's own theory, incorporating the 'flying gravel bank' idea, is that the material is not bound together by any substance. Rather a comet consists of a vast, irregularly shaped, swarm of tiny particles when it is at a great distance from the Sun. Each particle travels in its own independent orbit with its companions under the gravitational influence of the Sun. The separation of these particles is such, in relation to their size, that the transparent nature of comets is evident. Indeed, the total mass of a comet is estimated to be of the order of 10^{14} kg, which may seem large, but, at this figure, about 10^{11} comets would be needed to make up a mass equal to that of the Earth. On this theory the nearer the comet approaches the Sun the more rapidly the particles collide with one another, shattering into a myriad of smaller particles accompanied by intense local heating. The latter causes the conversion of some of the material to gases at high temperature. The gases can then emit light to give the spectra from which the constituents can be deduced. The self-gravitation of the larger particles allows them to stay together but the smaller dust particles are driven away from the coma to form part of the tail.

Fred Whipple's theory allows for the concentration of material at the centre of the coma; Professor Lyttleton's theory does not. The former theory tends to explain how long-period comets survive during their journey to great distances from the Sun by the process of producing an ice-bound conglomerate. The latter allows an explanation of the deviation of some comets from the theoretical orbits calculated solely on Newton's law of gravitation, whereas the former theory has to invoke arbitrary non-gravitational forces.

In spite of Professor Lyttleton's strong arguments against the 'dirty-snowball' model, many astronomers favour it as the most reasonable. The fairly recent discovery of water (H_2O) molecules in a comet (Kohoutek again, in 1973) gives support to the 'snowball' theory.

One of the prime objectives of the space missions to Halley's comet is to find out if a nucleus does exist and, if so, to photograph it. A further objective is to gain information on the distribution and size of the dust. If the missions are successful one or other theory, or neither, may be shown to be correct. For the time being we must sit on the fence.

19　Atoms, molecules, ions and isotopes

In replying to the question 'What is a comet?' (section 18) we refer to atoms, molecules, ions and isotopes. It may be helpful to give a brief explanation of these here.

The chemistry of the author's schooldays used to tell us that if we take a small piece of material such as gold and continue to divide and subdivide it by cutting, there will come a stage when we can divide it no more. This is not due to the limitation of the cutting process but rather to the fact that we have arrived at the stage where each tiny piece, now about one-hundred-millionth of a centimetre (10^{-8} cm) in size, is an *atom*. If all the atoms are exactly alike, we say that we have been dealing with the *element* gold. The same ideas can be applied to gaseous substances such as hydrogen. There are 92 elements which occur naturally, but more can be made by artificial processes. One atom of an element is usually designated by suitable letters, H for hydrogen, C for carbon, Na (natrium) for sodium and so on.

The modern view of the atom is much more complex, for atoms *can* be subdivided. Each atom has a central core, the *nucleus*, of size about 10^{-12} cm. The nucleus itself consists of (a) *protons*, each with a mass of about 1.7×10^{-24} g and carrying a *positive* electrical charge of about 1.7×10^{-19} coulomb, and (b) *neutrons*, each with a mass of about 1.6×10^{-24} g but carrying no electrical charge. The nucleus is thus charged positively overall.

One or more *electrons*, each with a mass of about 9×10^{-28} g and each carrying a *negative* electrical charge equal in size to the positive charge on the proton, revolve in orbits round the nucleus. The distances of these orbits from the nucleus are not casual but are generally predetermined within any atom. Also, an orbit may contain more than one electron. The total number of electrons usually equals the total number of protons, and the atom is neutral.

All atoms of the same element always contain the same number of protons but it is possible for some atoms of the same element to contain more neutrons than protons, in which case we say that we have an *isotope* of the element. Thus the appearance of C, ^{12}C and ^{13}C in comets indicates the presence of neutral carbon C and the two isotopes of carbon ^{12}C, ^{13}C, where the superscripts refer to the total number of protons and neutrons in an atom of the respective isotopes.

Where two or more atoms of the same element become chemically joined they form a *molecule*, for example C_2 in comets. If the atoms joined together are of different elements they form a *compound*, such as CN, the cyanogen in Comet Halley 1910. If, however, atoms of different elements joined together behave in chemical reactions as if they were a single atom they are known as *radicals*, an example in comets being the hydroxyl radical OH.

Radiation from the Sun falling on some molecules or radicals in the tail of a comet can cause them to lose an electron, leaving the molecule or radical positively charged or *ionized*. The molecules or radicals are written CO^+, OH^+ to indicate the positively charged nature of the *ions*.

The modern theory of the structure of the atom is very complex, but the outline given above should be sufficient for our purposes.

20   Where do comets come from?

As with the structure of comets (section 18), we seem to have two main theories from which to choose. They are that comets are left over from the time when the solar system was formed some 5 thousand million years ago, or that comets are formed from the dust between the Sun and neighbouring stars. In either case it seems that interstellar material is involved.

In 1950 the Dutch astronomer J. Oort envisaged that the

comets with long periods were formed in the vicinity of Jupiter's orbit or alternatively in the region beyond Neptune. They were then flung out into space either by the gravitational effects of the massive planet Jupiter or by an excessive solar wind from the Sun at a particular stage of its evolution. This idea leads to the postulation of a vast reservoir of cometary material equivalent to about 10^{11} comets, of total mass approximately equal to that of the Earth. This cloud of cometary material extends roughly from 40 000 a.u. to 100 000 a.u., enveloping the entire solar system. It includes a large number of comets orbiting the Sun at these great distances. The distance of the nearest star is just over 270 000 a.u. Disturbances in this 'Oort cloud' by the gravitational effects of a moving nearby star cause some of the cometary material to move away from the solar system where it is lost for ever, but a small fraction of this large assembly of matter is caused to move *towards* the solar system, reducing the size of the orbits of some of the comets.

Although the stars seem fixed in their relative positions in the sky, the movement of some of them can be detected on photographic plates taken some time apart. From the movement of these stars we can deduce that our Sun is also moving through space, at a speed of around 19 km/s. The British astronomer Raymond Lyttleton calls upon this movement of the Sun through the interstellar dust to account for the formation of comets.

If we were to travel with the Sun the interstellar dust would appear to be approaching us. The gravitational attraction of the Sun on some of this dust causes it to be deflected towards the Sun, as show in figure 20 by the lines AA' to arrive on the axis XX'. The Sun is acting as a gravitational 'lens' and 'focusing' the dust. On its arrival at the preferred axis XX', the dust has a component of its velocity across XX', but due to collisions between the arriving particles this component is destroyed and the

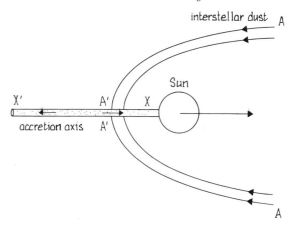

FIGURE 20   The formation of comets by gravitational focusing by the Sun.

particles stay within a cylinder with XX' as its axis, behind the Sun. The radius of this cylinder is estimated as about 100 000 km. The inflow of new material causes the material already in the cylinder to flow along the axis XX'. Those particles which are further from the Sun than say 150 a.u. flow away from the Sun, while those which are nearer flow towards the Sun. On this inward journey towards the Sun some of the particles come together to form comets of mass of about 10^{14} kg. This process takes place over a very long time-scale, so we should not expect showers of comets.

The mathematics supporting the latter theory of the formation of comets, of which Halley's comet might have been one, is ingenious and convincing, and by inserting reasonable values for his assumed quantities Professor Lyttleton arrives at values which correspond with observation. Furthermore he puts forward a strong opinion that the Oort shell of comets does not exist. Nevertheless some recent research by P. Weissman of the Jet Propulsion Laboratory, presented in a paper at the International Astronomical Union Colloquium 61 on comets, using the new and powerful tool of scientists, the large computer, indicates that the Oort theory is still very plausible. His study simulated the changes in the orbits of a wide range of comets with *aphelia* (points of greatest distance from the Sun) within the Oort cloud. Comets with aphelia of 40 000 a.u. and perihelia in the lower part of the range 20 to 10 000 a.u. move closer to the Sun when disturbed by other stars; those with perihelia in the upper part of the range move away out of the Sun's sphere of influence.

21   Brightness and magnitude

When we write of the *magnitude* of a comet we are referring to its brightness and not to its physical size. The *apparent brightness* of a comet, or a star, is the amount of energy, in the form of light, received by the eye of the observer in a given time. Astronomers today have instruments, such as the photoelectric photometer, which can measure this energy, but the human eye alone is a bad judge of brightness simply because it does not behave *linearly*. Equal increases of the light energy received by the eye do not result in the perception of equal increases in brightness. The response of the eye is *logarithmic*. This is one reason why the eye can cope with such a great range of brightness.

The *magnitude* scale is appreciated more by the eye. Most of the brightest stars which we can see with the unaided eye have a magnitude of about 1; most of the faintest stars just visible to the naked eye have a magnitude of about 6. This increase in magnitude of 5 corresponds to a hundred-fold *decrease* in brightness. The eye can easily distinguish a difference of one magnitude corresponding to a factor of 2.5 in brightness, and astronomers even estimate fractions of a magnitude, so the eye can readily work on this scale.

The mathematical relationship between brightness and magnitude is

$$2.5 \log_{10}(B_1/B_2) = m_2 - m_1$$

where B_1, B_2 are the brightness of two stars and m_1, m_2 are the corresponding magnitudes. If m_2 has a value of 6, less bright than m_1 of, say, value 1, then the formula gives $B_1 = 100 \times B_2$.

The magnitude scale can be extended in each direction. A negative magnitude is very bright (-12 for the Moon), and a large positive magnitude is very faint ($+24$ for Halley's comet at rediscovery in October 1982).

The different effects on the eye of equal increases in brightness and equal increases in magnitude can readily be demonstrated with the simplest of apparatus described in one of my previous books (*Projects and Demonstrations in Astronomy*, 1979).

We have used stars for our examples of magnitude but here, of course, we are interested in the magnitude of Halley's comet. Comets have two main parts, the head or coma and the tail. Two magnitudes are sometimes quoted for a comet: the magnitude of the coma alone which is usually the brightest part of the comet, and the *total* magnitude of coma and tail. In this book when we refer to the magnitude of Halley's comet we shall always refer to its total magnitude.

Because of the diffuse nature of a comet it is sometimes difficult to decide what its total magnitude is. One method is to compare the brightness of the comet with those of two nearby stars whose magnitudes are known. To make the stars look more like the comet we can put the telescope or binoculars slightly out of focus and then more readily estimate the magnitude of the comet in relation to the star images.

22   The brightness of comets

As a periodic comet approaches and then recedes from the Sun its brightness is not wholly predictable. The light from the comet consists partly of sunlight reflected from dust particles and partly from the fluorescence of atoms, molecules and ions caused by ultraviolet radiation from the Sun. The development of the head and of the tail are different at each reappearance. There is ample evidence (section 16) that the brightness of Halley's comet has varied enormously from one appearance to the next. The following paragraphs give an outline of a method used to predict the brightness of a returning comet. (Readers wishing to skip this more complex material may proceed to section 23.)

If the only light coming from a comet were reflected sunlight, the true brightness of the comet would be inversely proportional to the square of its distance r from the Sun, and its apparent brightness B to an observer on the Earth would be inversely proportional to the square of its distance Δ from the Earth.

Thus the expression for B would be:

$$B = B_0/(r^2 . \Delta^2)$$

where B_0 is the brightness which the comet would have as observed from Earth if the comet's distance from the Sun were 1 a.u. and its distance from the Earth were also 1 a.u. Of course it would be pure coincidence if the comet *were* 1 a.u. from the Earth at the same time as its distance from the Sun was 1 a.u., but this defines the constant of proportionality B_0.

But reflected sunlight is not the only light coming from the comet; each comet is different and each apparition of a particular comet is different from the ones before it. We therefore make the expression more general:

$$B = B_0/(r^n . \Delta^2) \qquad (1)$$

where B_0 and n have to be determined from observations made at the previous apparition or as the comet approaches. n typically has a value of say between 3 and 8, although it can be larger.

It is usual to think of brightness in terms of magnitude. The relation between brightness B, B_0 and the corresponding magnitudes m, m_0 is (see section 21):

$$2.5 \log_{10}(B_0/B) = m - m_0. \qquad (2)$$

Substituting $(B_0/B) = r^n . \Delta^2$ from (1) in (2), we have:

$$2.5 \log_{10}(r^n . \Delta^2) = m - m_0$$

or:

$$(2.5n) \log_{10}r + 5 \log_{10}\Delta = m - m_0$$

giving the apparent magnitude:

$$m = m_0 + 5 \log_{10}\Delta + (2.5n) \log_{10}r. \qquad (3)$$

We must recall that we do not have any values for m_0 and n. If we have not seen the comet on its return we must resort to the values obtained for m and the corresponding Δ, r at times during the previous appearance. If the comet *has* been seen on its return we can similarly use observed values of m and the corresponding Δ, r.

Re-grouping (3) we may write:

$$(m - 5 \log_{10}\Delta) = (2.5n) . \log_{10}r + m_0.$$

This has the form

$$y = a.x + b.$$

Points may be plotted on a graph with values of $(m - 5 \log_{10}\Delta)$ as ordinates and $\log_{10}r$ as abscissae.

If the foregoing theory is not wildly astray the points

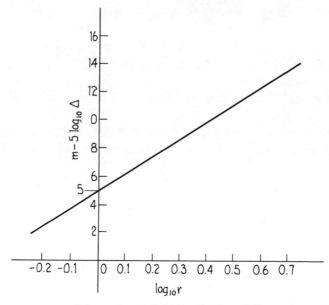

Halley's Comet before perihelion 1910

FIGURE 21 *Predicting the brightness of Halley's comet.*

should approximate to a straight line, and the 'best' straight line may be drawn using the method of 'least squares'. Using the straight line, the value of $(m-5 \log_{10}\Delta)$ when $\log_{10}r = 0$ will give the value m_0, and the slope of the straight line will give the value of $(2.5n)$.

Figure 21 gives the approximate position of the graph for the last appearance of Halley's comet in 1910 for the period *before* perihelion passage. The information in the figure has been reconstructed from that given by Donald Yeomans, who gives the value of $m_0 = 5.0$ and $(2.5n) = 13.1$.

So for this (1986) apparition the predictions for total apparent magnitudes before perihelion on 9 February 1986 are:

$$m = 5.0 + 5 \log_{10}\Delta + 13.1 \log_{10}r.$$

As an example, on 8 August 1985, if the predicted orbit is correct, we have Δ = 3.64 a.u. and r = 3.02 a.u. and the predicted total apparent magnitude is:

$$m = 5.0 + (5 \times 0.5611) + (13.1 \times 0.4800) = 14.1.$$

The predictions for Halley's comet *after* perihelion are not so reasonably secured because the graph for this period of the 1910 apparition did not approximate to a straight line. Donald Yeomans has nevertheless given predictions of total apparent magnitude after perihelion for the 1986 apparition using a special technique of fitting an equation to the 1910 curve.

23   Spectroscopy of comets

When a narrow beam of white light is passed through a transparent prism it emerges as a band of colours which change from red (long wavelength) at one end to violet (short wavelength) at the other through the colours of the rainbow. This band of colours is called a *continuous spectrum* (figure 22(a)).

We get much the same sort of spectrum from the prism if the light entering it comes from a hot glowing body, irrespective of what the body is made of. If, however, the body is heated until it vaporizes and the light from the hot vapour is passed through the prism, the band of colours is weakened but it is crossed by a series of bright lines. Each element in the vapour provides one or more of the bright lines, each in a definite position relative to the others. It is, therefore, possible to tell by measurement of the spectrum which elements are present in the vapour. The same is true of a glowing rarefied gas. Such a spectrum is said to produce *emission lines*.

These lines arise from the fact that electrons orbit the nucleus of an atom of an element of the gas at distances

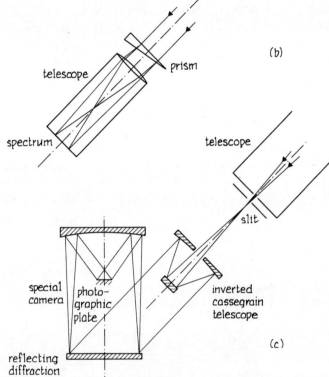

FIGURE 22   (a) The formation of a continuous spectrum.   (b) A simple
spectrograph using a prism and a telescope.   (c) A slit spectrograph for a
large telescope incorporating a diffraction grating and a camera.

corresponding to their energies. If an electron or other particle coming from the Sun collides with an atom of the gas of a comet the energy of an electron of the gas may increase so that the electron moves into an orbit of higher energy level. The situation soon normalizes, with the electron changing its orbit to the one of lower level so that it can now give out a packet of energy in the form of light (a *photon*) of a particular wavelength. This is detected as an emission line by a spectrograph and then photographed at the eyepiece end of the observing telescope. At the same time, light of known wavelengths is fed into the spectrograph so that the lines it produces can be used to calibrate the spectrum of the comet (see plate 15).

One of the simplest types of spectrograph is shown in figure 22(b). The prism is placed in front of the objective lens (furthest from the eye) of the telescope and, because the rays of light coming from the comet are virtually parallel, the spectrum is formed at the eyepiece where it may be photographed. Simple in principle, this method has the disadvantages: (a) that the light coming to the objective lens not only from the comet but also from nearby stars forms multiple images at the photographic plate: (b) that the spread of the spectrum is not great.

Instead of a prism we may use a *diffraction grating*, which has the same effect but has the advantage of giving a brighter and more widely spread spectrum. A diffraction grating consists of a flat mirror on which a number of close parallel lines are ruled. Light may pass through the rulings or it may be reflected at the mirror rulings left untouched. Rulings of 600 lines/mm are not uncommon.

There is a single slit in the telescope so that if the telescope is directed in such a way that the light from the comet alone is coming through the slit, only the spectrum of the comet will be recorded. A slit of width of the order of 0.1 mm may be used. A slit spectrograph which can be attached to a telescope is shown in figure 22(c).

The spectrum of Halley's comet 1910 is shown in plate 15, on which emission lines can be clearly seen. The bright lines are formed as described above and are superimposed on a continuous spectrum formed by sunlight reflected from the dust particles of the comet.

24   The velocity of Halley's comet

(This section may be skipped if desired. Readers may proceed to section 25.) The formula for the velocity v of Halley's comet in its orbital path when it is at a distance r from the Sun is derived from the general formula for the velocity of any body moving in an orbit round the Sun under the gravitational influence of the latter alone. The proviso is made that the orbiting body has a mass which is negligible compared with that of the Sun. This is clearly the case for the motion of comets. The formula is

$$v^2 = G.M(2/r - 1/a)$$

where G is called the universal gravitational constant, M is the mass of the Sun, a is the semi-major axis of the orbit. The values, expressed in metres, kilograms and second units, are:

$$G = 6.67 \times 10^{-11} \text{ m}^3 \text{ kg}^{-1} \text{ s}^{-2}$$

$$M = 1.99 \times 10^{30} \text{ kg}$$

$$a = 17.94 \times (150 \times 10^6 \times 10^3) \text{ m.}$$

However, it is convenient to work with the astronomical unit instead of the metre which gives, for Halley's comet:

$$G = \frac{6.67 \times 10^{-11}}{(150 \times 10^6 \times 10^3)^3} \text{ a.u.}^3 \text{ kg}^{-1} \text{ s}^{-2}$$

$$M = 1.99 \times 10^{30} \text{ kg}$$

$a = 17.94$ a.u.

The velocity v will then initially be calculated in a.u./s. Thus:

$$v^2 = \frac{6.67 \times 10^{-11}}{(150 \times 10^6 \times 10^3)^3} \times 1.99 \times 10^{30} \ (2/r - 1/a)$$

$$v^2 = 0.000003933 \times 10^{-8} \ (2/r - 1/a)$$

$$v = 0.00198 \times 10^{-4} \ \sqrt{(2/r - 1/a)}.$$

Changing to kilometres per second, for Halley's comet:

$$v = 0.00198 \times 10^{-4} \times (150 \times 10^6) \times \sqrt{(2/r - 1/17.94)}$$

or,

$$v = 29.7 \ \sqrt{(2/r - 0.0557)}$$

where r is in a.u.

25 Demonstrating the relative motions of Halley's comet and the Earth: computer programs

The following program, HALLEY, illustrates in truly equal time intervals on a black sky the motion of Halley's comet and the Earth relative to the Sun from July 1985, when Giotto is launched, to July 1986.

For explanatory notes on all the computer programs see appendix A, and for a version of this program for the ZX81, see appendix B.

The values given in table 2 below will provide a realistic demonstration.

TABLE 2

	Computer input	Comet Halley
Gaussian constant	K	6.28 a.u. per year
Perihelion distance	Q	0.59 a.u.
Year of perihelion passage	P	1986
Month of perihelion passage	M	2.3 approx. 9 February
Year of initial date	S	1985
Month of initial date	M	7.3 approx. 10 July
Time interval	W	1 month
Number of dates	U	13

BBC Model B

```
 10 REM "HALLEY"
 20 MODE 4
 30 PRINT "HALLEY,EARTH,SUN DEMONSTRATION"
 40 PRINT
 50 DIM A(40):DIM B(40):DIM F(40):DIM L(40)
 60 DIM C(40):DIM D(40):DIM H(40)
 70 DIM M(40):DIM N(40):DIM R(40)
 80 DIM X(40):DIM Y(40)
 90 INPUT "Gaussian constant (AU/year)",K
100 INPUT "Perihelion distance ",Q
110 PRINT
120 INPUT "Year of perihelion passage ",P
130 REPEAT INPUT "Month of perihelion passage ",M:UNTIL M>=1 AND M<13
140 P=INT(P)+(M-1)/12
150 INPUT "Year of initial date ",S
160 REPEAT INPUT "Month of initial date ",M:UNTIL M>=1 AND M<13
170 S=INT(S)+(M-1)/12
180 INPUT "Time interval (months) ",W
190 W=W/12
200 REPEAT INPUT "Number of dates (5-40) ",U: UNTIL U>=5 AND U<=40
210 PRINT "Please wait..."
220 FOR I = 1 TO U
230 LET M(I)=K*(S+(I-1)*W-P)/(SQR(2*Q^3))
240 IF I = 1 THEN LET A(I)=M(I)
250 LET H(I)=-(A(I)*A(I)*A(I)+3*A(I)-3*M(I))/(3*A(I)*A(I)+3)
260 LET B(I)=A(I)+H(I)
270 LET C(I)=B(I)*B(I)*B(I)+3*B(I)-3*M(I)
280 LET D(I)=ABS C(I)
290 IF D(I)<0.0001 THEN GOTO 320
300 LET A(I)=B(I)
310 GOTO 250
320 LET N(I)=ATN B(I)
330 NEXT I
340 FOR I = 1 TO U
350 LET R(I)=Q*(1+TAN N(I)*TAN N(I))
360 LET N(I)=2*N(I)
370 LET X(I)=150*R(I)*COS N(I)
380 LET Y(I)=150*R(I)*SIN N(I)
390 LET F(I)=150*COS((I-2)*PI/6)
400 LET L(I)=150*SIN((I-2)*PI/6)
410 NEXT I
420 MODE 1
430 VDU 5
440 VDU 23,241,0,0,0,0,0,0,8,0
450 MOVE 700,550
```

```
460 GCOL 0,1
470 PRINT CHR$(43)
480 FOR I = 1 TO U
490 MOVE 700+X(I),550-Y(I)
500 GCOL 0,2
510 PRINT CHR$(44)
520 MOVE 700+F(I),550+L(I)
530 GCOL 0,3
540 VDU 46
550 FOR T = 1 TO 1500:NEXT T
560 VDU 127:VDU 241
570 NEXT I
580 VDU 4
590 END
```

Spectrum

```
 10 REM "HALLEY"
 20 PRINT "HALLEY,EARTH,SUN DEMO"
 30 PRINT
 40 DIM a(40): DIM b(40): DIM c(40)
 50 DIM d(40): DIM h(40): DIM m(40)
 60 DIM n(40): DIM r(40)
 70 DIM x(40): DIM y(40)
 80 INPUT "Gaussian constant (AU/year) ",k
 90 INPUT "Perihelion distance (AU) ",q
100 PRINT
110 INPUT "Year of perihelion passage ",p
120 INPUT "Month of perihelion passage ",m
130 IF m<1 OR m>12.99 THEN GO TO 120
140 LET p=INT(p)+(m-1)/12
150 INPUT "Year of initial date ",s
160 INPUT "Month of initial date ",m
170 IF m<1 OR m>12.99 THEN GO TO 160
180 LET s=INT (s)+(m-1)/12
190 INPUT "Time interval (months) ",w
200 LET w=w/12
210 INPUT "Number of dates (5-40) ",u
220 IF u<5 OR u>40 THEN GO TO 210
230 PRINT "Please wait..."
240 FOR i=1 TO u
250 LET m(i)=k*(s+(i-1)*w-p)/(SQR (2*q^3))
260 IF i=1 THEN LET a(i)=m(i)
270 LET h(i)=-(a(i)*a(i)*a(i)+3*a(i)-3*m(i))/(3*a(i)*a(i)+3)
280 LET b(i)=a(i)+h(i)
290 LET c(i)=b(i)*b(i)*b(i)+3*b(i)-3*m(i)
300 LET d(i)=ABS c(i)
310 IF d(i)<0.0001 THEN GO TO 340
320 LET a(i)=b(i)
330 GO TO 270
340 LET n(i)=ATN b(i)
350 NEXT i
360 FOR i=1 TO u
370 LET r(i)=q*(1+TAN n(i)*TAN n(i))
380 LET n(i)=2*n(i)
390 LET x(i)=4*r(i)*COS n(i)
400 LET y(i)=4*r(i)*SIN n(i)
410 LET x(i)=20-q+x(i)
420 LET y(i)=10+y(i)
430 NEXT i
440 PAPER 0: INK 5
450 CLS : PRINT AT 10,20; CHR$ (42)
460 FOR i=1 TO u
470 PAPER 0: INK 6
480 PRINT AT INT (y(i)+0.5),INT (x(i)+0.5); CHR$(44)
490 PAPER 0: INK 2
500 PRINT AT 10-4*(SIN ((i-2)*PI/6),20+4*COS ((i-2)*PI/6); CHR$(43)
510 PAUSE 50
515 PRINT AT 10-4*(SIN ((i-2)*PI/6),20+4*COS ((i-2)*PI/6); CHR$(32)
520 NEXT i
```

26 A three-dimensional model of the orbit of Halley's comet

Figure 5 on page 37 shows the orbit of Halley's comet as if it were in the same plane as the plane of the orbit of the Earth round the Sun. It is, however, inclined at about 18 degrees to the plane of the Earth's orbit. This is shown diagrammatically in figure 23 (albeit still on two-dimensional paper).

We shall now describe how a simple three-dimensional model can be made, using only two pieces of cardboard, to illustrate the space relationships between the comet, the Earth and the Sun for the period when the comet is near both Earth and Sun. It will also give us an appreciation of what is meant by the *elements* of the orbit.

Two approximations will be made, both of which are not far from the truth to the scale used: first, the orbit of the earth E is a circle with the Sun S at the centre; second, the orbit of Halley's comet is very nearly parabolic for the short time considered.

Figure 24(i) illustrates the position of γ the First Point of Aries, the angles Ω, ω, i and q, the perihelion distance, all of which are used in the construction of the model. On one piece of card we construct the orbit of the Earth; on the other piece of card we construct the orbit of Halley's comet. Then we interlock them at the appropriate angle.

The values for Halley's comet are:

Longitude of ascending node	Ω	$= 58°$
Longitude of perihelion	ω	$= 112°$
Inclination of orbit	i	$= 162°$
Perihelion distance	q	$= 0.59$ a.u.

On a piece of fairly stiff card 150 mm square draw a circle of radius 25 mm with its centre located as in figure 24(ii). This represents the orbit of the Earth E round the Sun S, so

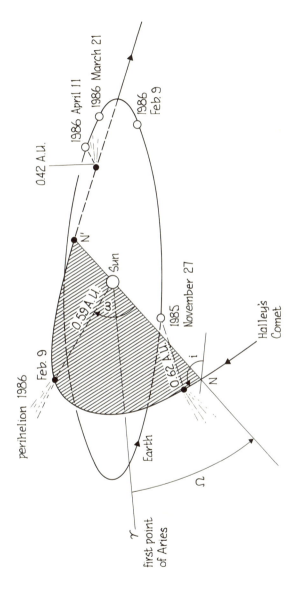

FIGURE 23   The relation between the planes of the orbits of Halley's comet and the Earth.

The Science of comets

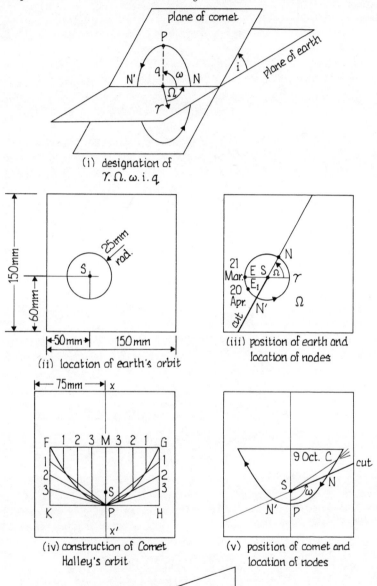

(i) designation of
Υ. Ω. ω. i. q

(ii) location of earth's orbit

(iii) position of earth and
location of nodes

(iv) construction of Comet
Halley's orbit

(v) position of comet and
location of nodes

(vi) triangle to set angle between planes

we have chosen a scale of 25 mm = 1 a.u. Referring to figure 24(iii), mark γ at the right-hand end of the horizontal diameter $ES\gamma$. γ represents the position of the Sun as seen from the Earth on 21 March each year so we may now label E with this date.

The Earth moves once round this circle (360°) in one year (365 days), which is very nearly one degree per day. If E_1 represents the position of the Earth on, say, 9 May, 49 days later, then E_1 can be marked on the circle by measuring angle ESE_1 49° *anticlockwise* from ES. Similarly, mark on the circle the position of the Earth for each date given in table 3.

Now draw a line $N'SN$ such that $\gamma SN = \Omega$, also measured anticlockwise from S. This is the line of intersection of the planes of the two orbits (see figure 24(i)). Cut the card in a straight line from its edge along the line $N'S$ as far as S *only*.

On the second piece of card 150 mm square we construct the orbit of Halley's comet to the same scale (25 mm = 1 a.u.). Referring to figure 24(iv), draw the centre line XX'. Locate the position of the Sun S, at the focus of the parabola, on XX' 50 mm from the edge X'. Mark the vertex P on the parabola such that $SP = q(=14.75$ mm). Set up a rectangle $FMGHPK$ with sides parallel to those of the card such that $PM = MF = MG = 4q$. Divide MG into four equal parts. Divide GH into four equal parts. Locate points on the parabolic orbit using the construction lines shown in figure 24(iv). Do the same for $MFKP$. Draw in the parabola through these points and the vertex.

To locate the position of the comet at any date in table 3 convert the distance r of the comet from the Sun for that

FIGURE 24 *(opposite)   Making a model of the orbits of Halley's comet and the Earth: (i) designation of γ, Ω, ω, i, q;   (ii) location of the Earth's orbit;   (iii) position of the Earth and location of the nodes;   (iv) construction of the comet's orbit;   (v) position of the comet and location of the nodes;   (vi) triangle to set the angle between the planes.*

date to our scale. Thus, on 9 October r is 2.24 a.u. (= 56 mm). With centre S and this radius strike an arc to cut the parabolic orbit in C (see figure 24(v)). Similarly locate the position of the comet for all other dates in table 3.

Draw the line of the nodes $N'SN$ such that $PSN = \omega$ measured *anticlockwise* from PS to cut the orbit in N' and N. The comet will be at position N, the ascending node, when it crosses the plane of the Earth's orbit from the south side to the north side, and at N', the descending node, when crossing from the north side to the south side. Cut in a straight line from the edge of the card and along NS as far as S *only*.

Push the two cards together so that the cuts engage as far as S. Seal the raw edge of that part of the cut which remains on the Earth card, using masking tape or sellotape. Colour that part of the comet's orbit which protrudes on the north side of the Earth's orbit.

Finally, we must fix the two planes at the correct angle. This may be done by cutting a triangle (figure 24 (vi)) with one angle 18° (180°−162°) and locating it *underneath* the Earth's orbit at the position of the Earth on about 9 February. From this model we can see:

(a) that on approach the comet is nearest the Earth on 27 November 1985;
(b) that on receding the comet is nearest the Earth on 11 April 1986);
(c) that at perihelion the comet and the Earth are on opposite sides of the Sun;
(d) the relative positions of the Earth, the comet and the Sun at Halley encounter by Giotto on 13 March 1986.

Clearly if other materials are available a more attractive and more rigid model can be made, based on the construction described.

TABLE 3
Distance *r* of Halley's comet from the Sun (a.u.)

Date		*r* (a.u.)	
1985	September 9	2.63	
	October 9	2.24	
	November 9	1.81	
	November 27	1.55	comet nearest Earth (1)
	December 9	1.36	
1986	January 9	0.89	
	February 9	0.59	comet nearest the Sun
	March 13	0.89	Giotto mission encounter
	April 11	1.33	comet nearest Earth (2)
	May 9	1.75	
	June 9	2.18	
	July 9	2.57	

IV

Anecdotes, Excerpts and an Important Theory

27 Anecdotes and excerpts

From the time of Isaac Newton and Edmond Halley it became possible, due to the work of both these eminent scientists, to forecast the reappearance of the comet which now bears the name of the latter, and of other comets too. Not only can the time of reappearance be accurately predicted, but science has given us details of some of the constituents of this comet (see section 18) and information regarding its coverage of the sky. This foreknowledge has sometimes been the cause of unwarranted concern, particularly around AD 1910.

Before we come to the 1910 apparition, about which there is the most information, it is as well to reflect on what might have been the course of the prediction of the return of comets, and indeed the whole of celestial mechanics, if Isaac Newton had not survived his birth. We may quote James R. Newman's 'Commentary on Isaac Newton' in *The World of Mathematics* (1956):

> On Christmas Day 1642, the year Galileo died, there was born in the Manor House of Woolsthorpe-by-Colsterworth a male infant so tiny that, as his mother told him in later years, he might have been put in a quart mug, and so frail that he had to wear a bolster around his neck to support his

head. This unfortunate creature was entered in the parish register as 'Isaac sonne of Isaac and Hanna Newton'.

There is no record that the wise men honoured the occasion, yet this child was to alter the thought and habit of the world.

Newton also survived the Great Plague of 1665, presumably because no comet deposited the organisms which carried it (see section 28) near the farm at Woolsthorpe to which Newton had temporarily retired (because Cambridge University was closed owing to its proximity to London where the plague did rage).

But what of Edmond Halley himself? By his own account he was born on 29 October 1656, in the relatively quiet political climate between two wars, in Shoreditch near London. No known parish register records the fact and no baptism certificate exists as far as we know.

It is recorded in the minutes of the Royal Society of London that this learned society generously authorized the printing of Newton's *Principia*, but were not prepared to fund it. Instead the Society passed the resolution 'that Mr Halley undertake the business of looking after it, and printing it at his own charge'. It is significant that in spite of Halley's commitment both in effort and in finance the person whose name appears on the title page giving the blessing of the Royal Society to the publication was in fact diarist Samuel Pepys, who happened to be the current President of the Society (see plate 2).

At the time at which 'Halley's' comet appeared in 1682, and on whose details Halley was to predict the return of 1758, he was also busy producing mortality tables connected with life insurance problems, a task which could not be much further away from the work for which he is best remembered.

The newspapers and magazines at the AD 1910 apparition carried much news and opinion about the reappearance of Halley's comet. The magazine *Punch* (or the *London*

Charivari) has a world-wide reputation for its own brand of humour. We quote the following extract without comment.

HOW TO SEE HALLEY'S COMET

Many earnest correspondents ask us to reassure them as to the great comet's visibility. They gather from the newspapers that the 'unique celestial display' will be difficult to find. . . . The following general directions will be valued by many other correspondents with whom we have no space to deal separately. To find Halley's comet stand so that Mars and Saturn are in line running N.N.E. by E. from the parallax, and making an angle of 34 degrees with an imaginary line drawn S.E. by E.S.E. (approximately) from the Pole Star, Venus or the constellation of Orion. Taking then the mean distance between any two of these points and Greenwich, draw a pentagonal focus-line directly opposite to the base of observation. This will give the required altitude, and the curve of contact will be instantly apparent.

More dramatically, the predicted path of Halley's comet prior to its 1910 appearance indicated that the Earth would pass through the comet's tail on 18 May 1910 if the tail were more than 15 million miles long. As mentioned in section 6, the development of spectroscopy since the 1835 apparition had allowed the identification of molecules of the poisonous gas cyanogen (CN) and a cloud of hydrogen gas in the head of another comet (Comet Morehouse 1908). Indeed, CN *was* also detected at the 1910 apparition of Halley's comet.

The *New York Times* reported calmly:

Tail 46 million miles long? – scarfed in a filmy bit of it, we'll whirl in our dance through space, unharmed, and most of us unheeding.

In the comet's tail on Wednesday – European and American astronomers agree the Earth will not suffer in the passage.

Six hours tonight in the comet's tail – few New Yorkers are likely to know it, for it may be cloudy.

and not so calmly:

> Chicago is terrified – women are stopping up doors and windows to keep out cyanogen.

and perhaps religiously:

> Balloon trip to view comet – Aeronaut Harmon invites College Deans to join him in Ascension.

It is reliably reported that pills for protection against the comet sold well and that in a town in the interior of Asia Minor, people prepared barrels of water in which to immerse themselves during the predicted period of passage of the Earth through Halley's tail, though there is no evidence that anybody did.

In the local newspapers of Wisconsin in 1910 there was a furniture advertisement offering 'seats for comet gazers'. There was a recipe for a 'Comet Cocktail' and a report that 'Comet parties are all the rage'.

Macleans magazine in 1955 carried 'a comprehensive account of the strange and wonderful behaviour provoked by fear of the comet in 1910'.

It is easy for us now with our advanced knowledge to scoff at all this, but perhaps one day a comet, but not Halley's, will strike the Earth with devastating results. Halley was concerned about such a possibility, but the nearest known approach of Halley's comet to the Earth was 3.7 million miles in AD 837. The Tungsaka devastation in Siberia in 1908 could possibly have been due to the impact of a comet or part of a comet. Furthermore, some scientists have advanced the theory that the extinction of the dinosaurs could be explained as the result of a comet striking the Earth.

Look again at the picture of part of the Bayeux Tapestry (plate 10) which, although commonly attributed to Queen Matilda, the wife of William of Normandy, was actually designed by English artists under the guidance of Odo,

Bishop of Bayeux, who is also shown taking part in the Battle of Hastings. Harold appears to be falling out of his chair at the news of the (AD 1066) apparition of the comet. In the scene shown, the ships below Harold are not filled in and may be taken to be 'ghost ships' in Harold's imagination before the invasion by William.

Halley ventured into the poetic with an Ode to Newton which, although not included in the manuscript, nevertheless appears in the first edition of the *Principia*. Two lines of the forty-eight-line Ode refer to Newton's vital work on comets. His reference to the comet as a *bearded* star corresponds with descriptions by the classical Romans of comets as 'hairy' stars.

Mark Twain was born in 1835, while Halley's comet was still visible, and commented that as he had come to the Earth with it, he would be greatly disappointed if he did not also go out with it. Indeed, after it had passed perihelion in 1910 Mark Twain died.

There is in England a Halley's Comet Society which meets once a year in venues such as Lords' Pavilion and the Royal Greenwich Observatory. Membership may be had for the price of a tie or medallion. They make it a rule to pronounce Halley's name as 'Hawley'.

Finally, the following is one of four verses reported in the Records of the Royal Astronomical Society Club to have been sung in 1910 to the tune of 'Sally in our alley'.

> Of all the meteors in the sky
> There's none like Comet Halley,
> We see it with the naked eye
> And periodically.
> The first to see it was not he,
> But still we call it Halley,
> The notion that it would return
> Was his originally.

28 Was life brought to Earth by comets, and do comets bring disease?

These two questions are not specifically linked to Halley's comet, but it could be as involved as any other comet. The Earth crosses the debris from Halley's comet in May and October each year.

The two questions themselves are linked as part of a theory put forward by Professors Sir Fred Hoyle and Chandra Wickramasinghe of University College, Cardiff. The former is an eminent astronomer noted for his ability to deviate from the accepted paths of science and to produce theories, from his original thinking, which are not always readily acceptable to his scientific colleagues.

In three books (see bibliography), these two scientists have pooled their specialist knowledge of astronomy and biology. They reject the accepted view that life on Earth began on Earth. Instead they take the view that it is entirely possible, even likely, that comets, which travel the great distances from the space between the stars (see section 20), collected the bacteria from which life evolved and deposited them on the Earth's surface.

The space between the stars contains vast quantities of interstellar dust. Sir Fred and his colleague argue that the size of the interstellar grains is similar to the size of bacteria from which life could have evolved. Some of the supporting evidence for this view is obtained from curves showing the extinction of light from stars as it passes through the dust. Two scientists in America suggest that the curves could more readily be explained if the dust were made up of the more theoretically acceptable silicon compounds and water-ice. Nevertheless, astronomers using radio telescopes have detected organic molecules in interstellar dust and gas.

Comets are known to contain molecules from which the

much more complex molecules necessary for life could evolve. Life chemistry is very complicated, but we cannot ignore the fact that the Apollo 11 and 12 space missions brought back traces of amino acids, essential to life, in rocks collected on the Moon. The Apollo 14 mission confirmed these findings. The amino acids could have arrived during the bombardment of the Moon in the early stages of the formation of the solar system. A report from Moscow told of the discovery in 1972 of a double spiral structure similar to DNA (a substance found in chromosomes of higher organisms, which stores genetic information) in a meteorite which landed in the Ukraine in 1889. All such Earth findings are suspect because of possible contamination. The results from the Moon are different in that the Moon is sterile.

The appearance of Comet Kohoutek in 1973 was disappointing astronomically because its actual brightness was much less than its predicted brightness. It did, however, provide us with two new important scientific achievements. It was the first comet to be studied from a manned artificial Earth satellite (Skylab), which gives promise for the future. Comet Kohoutek was found to contain water molecules, essential for life-forming processes.

Returning to the fundamental question of whether life could have arrived on Earth from comets, our two professors have provided answers for the awkward subsidiary questions of how the bacteria could survive the extreme cold of interstellar space and exposure to ultraviolet radiation. They even suggest the mechanisms by which bacteria could fall to the Earth from their high velocities of approach in cometary material without being burned up in the Earth's atmosphere. On Earth life survives the most rigorous environmental conditions. Some bacteria and yeast cells can withstand long periods at a temperature near absolute zero ($-273°C$), the intense cold of outer space, and they can even grow at $-20°C$. At the other extreme, some bacteria seem unaffected in acid hot springs at nearly 100°C, the temperature at which water boils. Seeds which have been frozen in sediments for thousands of years still grow. Bacteria which repair DNA after being exposed to radiation 5000 times that which will kill a person are known. Other forms of life, algae and fungi, seem unaffected by chemical conditions hostile to man himself. Life will survive at pressures of 1000 times that of our atmosphere and in a high vacuum. But life does not form without water, and this has now been discovered to be a constituent of comets. Tantalizing, yes; possible, yes; but the weight of scientific opinion is against life transmitted to Earth via comets. There is, however, recognition that the chemicals necessary for the formation of life could have arrived from space. Again, the 1986 space probes could help to resolve this fascinating problem.

Overall, the professors have provided an unorthodox, but well-reasoned, proposition of how life could have arrived on earth from comets. Much of the scientific world remains sceptical, but it is significant that Hoyle and Wickramasinghe are not the only ones considering this question seriously. In 1980 Cyril Ponnamperuma, a distinguished biochemist at the University of Maryland, USA, organized a meeting of about 100 astronomers and biochemists to consider whether there is a scientific basis for the idea of life on our planet being seeded by a comet. In 1983 Fellows of the Royal Astronomical Society devoted a whole morning to discussing the theory. That such a large number of scientists gave up time to this question indicates that Sir Fred must, as always, be taken seriously. A number of laboratory experiments set up to reproduce conditions similar to comet chemistry were reported, chiefly involving the irradiation of various molecules under different conditions. The experiments produced some unexpected results but were tantalizingly inconclusive.

But what of the second main question? Do comets bring diseases? Much of what has been said above about the transportation of bacteria from interstellar space to the environment of the Earth applies equally to disease-bearing viruses. Professors Hoyle and Wickramasinghe suggest that conditions exist in comets not only for the transportation but also for the rapid multiplication of bacteria and viruses. Important conclusions included in their opinions are that epidemics and pandemics of illnesses such as influenza and the bubonic plague do not occur because of person-to-person contact. Once the viruses arrive in cometary material in our upper atmosphere their falling and spreading to places on the surface of the Earth is very much dependent on the natural air currents. Descent into the stratosphere may take place in a matter of hours but the fall from that level to the Earth's surface may take years. The professors give evidence (not their own) of the dependence of the incidence of the influenza epidemic of 1950–51 on the global air currents. On a more local scale, their own researches on influenza in boarding schools in England, where there are accurate records and the topography is well-known, lead them to the same conclusions.

The spread of the bubonic plague in Europe in 1348 is shown to be along contours which do not correspond with the normal travel routes. Again, more locally, possibly the deposition of the plague virus of 1665 in the village of Eyam in Derbyshire is an example. It is said that the plague arrived in Eyam in bale of a cloth from London, but the villagers then isolated themselves to avoid spreading the disease. If Sir Fred and Chandra Wickramasinghe are right they did this needlessly. At least 250 villagers died – but about half this number did not, in spite of the close personal contacts (see plate 16).

Whether the above theories are correct or not, Sir Fred Hoyle can be relied upon to stimulate others to think.

PLATE *16   The stone on the boundary of Eyam, Derbyshire. During the Great Plague of 1665 the holes were filled with vinegar and coins dropped in to pay for provisions from neighbouring villages in an attempt to prevent the disease from spreading.*

V

Professional and Amateur Astronomers

29   What are we doing about observing Halley's comet?

It is worth recording here again that Halley's comet was found, or *recovered*, on 16 October 1982, by D. C. Jewitt and G. E. Danielson assisted by a few others, after an absence of 71 years. They used the charge-coupled device (CCD) (see section 31) of the Space Telescope Wide-Field Investigation Definition Team at the prime focus (see figure 27) of the 200-inch telescope at Mount Palomar. Comet Halley was then only 6 seconds of arc from the position predicted by calculation, near the star Procyon. Its magnitude was slightly fainter than 24, its distance was 1635 million km and it was moving about 3.5 seconds of arc in an hour.

These details were soon confirmed on 18 and 20 October 1982 by M. J. S. Bolton and H. Butcher at the Kitt Peak National Observatory, Arizona, and official recognition of the recovery was quickly given by the Central Bureau (see appendix C).

The positions derived from these observations indicate that Halley's comet will be nearest the Sun at about 7 o'clock in the morning of 9 February 1986, as predicted by Donald Yeomans.

The interest of amateur and professional astronomers

and of the layman will intensify as Comet Halley comes nearer to the Earth. The motion of the comet, the structure of its head and tail and its material content will be investigated by all possible means.

Ground-based observations from Earth will be coordinated under a project known as International Halley Watch (IHW), operated from the Jet Propulsion Laboratory in Pasadena, California. Specialists for the various aspects of comet science have been selected and an Amateur Observation Net set up. The IHW have issued an Observer's Handbook and a leaflet entitled *Seeing Halley's Comet*. Two other organizations, based in the United Kingdom, will collect information on Halley's comet and cooperate with the IHW to ensure maximum benefit. The Comet Section of the British Astronomical Association, with its many experienced comet observers, has its own programme for collecting visual, photographic, spectrographic and photo-electric observations. The BAA is represented on the Comet Halley UK Coordinating Committee (CHUKCC), as are the Royal Astronomical Society, the Royal Observatories (at Herstmonceux and Edinburgh), Jodrell Bank, the Rutherford Appleton Laboratory at Oxford, the Science and Engineering Research Council, the British Antarctic Survey and university teams working with both ground-based and satellite/probe facilities.

Although Halley's comet will be observed as often as possible, Halley Watch Days will be designated so that cooperative efforts can be made at the same times, especially during the period 8–14 March 1986 when the spaceprobe (section 30) will be near P/Halley.

To obtain experience of the observing techniques and procedures for reporting on Halley's comet, trial runs have been carried out on another periodic comet, P/Crommelin (1983n), which returns every 27 years, and on this occasion was at perihelion in February 1984. British astronomers also used the International Ultraviolet Explorer (IUE) satellite to make ultraviolet observations of P/Crommelin.

The theme for National Astronomy Week in the UK, 9–16 November 1985, will be Edmond Halley and Halley's comet. These dates have been chosen because the comet will then be well placed for evening observation and the Moon will not be prominent (section 15).

The most ambitious attempts to gain scientific information on Comet Halley will undoubtedly be made from space probes which will be sent to fly by or rendezvous with the comet. The Japanese will launch two such spacecraft, MS-T5 and Planet A; the Russians will use two *Venera* space vehicles; and the European Space Agency will send a sophisticated space probe Giotto to come within 500 km of the nucleus of Halley's comet. Details of all these missions are given in section 30.

With its most successful space programme, involving excellent data collection and accurate targeting to the Moon, Mercury, Venus, Jupiter and Saturn, the United States surely must have been the favourite for such a mission. However, because of other priorities the US government has decided not to allocate funds to launch a spacecraft specifically to fly by or rendezvous with Halley's comet. Other schemes using missions designed for other purposes by NASA have been studied, with a view to collecting what information they can relevant to Halley's comet.

To this end the United States has deflected an existing spacecraft, the International Sun–Earth Explorer (ISEE 3), using the Moon's gravitational force, into a trajectory which could swing the ISEE 3 to within 15000 km of another periodic comet, P/Giacobini-Zinner. In achieving this and other objectives, a most complex series of manoeuvres had to be undertaken, the last one in December 1983. The final trajectory past the Moon took ISEE 3 only 100 km above the Moon's surface and sent the spacecraft towards comet P/Giacobini-Zinner, where it will study the mechanism of the release of cometary material and also the composition of the comet. The instruments already on board the spacecraft will not allow any photography but the information gained about the hazards to be expected, particularly those caused by cometary dust, may be useful for the other missions. The ISEE 3 encounter with its comet will not occur until September 1985, after the launch of the other spaceprobes, but there is flexibility for modification of the Giotto programme even after launch.

The coronagraph of the Solar Maximum satellite, recently repaired by the space shuttle crew, may be used to photograph Halley's comet when the latter is too close to the Sun for Earth-bound observation.

There will be an international cooperative programme through the Inter-Agency Consultative Group to make the most of the scientific returns. The European, Soviet and Japanese space agencies as well as NASA are all members of the group. The British Interplanetary Society will be constantly updating its information about Halley's comet.

Possible radar observations of Halley's comet

We are probably all familiar with the basic principles of radar, which was used so successfully in the 1939–45 war. Radar equipment is used universally today on aircraft and on most large ships. Without it, even crossing the English Channel by boat can be a hazardous business because of the overcrowded shipping lanes. Essentially, a pulse of electrical energy is sent by a transmitter towards the target, and, after reflection from the target, some of the energy is received back at the site of the transmitter and displayed on a screen similar to a television tube. From the position and appearance of the display of the reflected signal on the tube the position and range of the target, and in some cases its appearance, can be deduced.

We must also be familiar with radio telescopes, for they can be seen dotted about the countryside. They look like very large saucers or 'dishes', as they are called, with a tripod spacing a small aerial from the centre of the dish. The best known of these must be one at Jodrell Bank, Cheshire, England. They are there to receive signals coming from distant astronomical objects and from the nearer artificial satellites circling the Earth. The dishes are parabolic in shape (conics again!). Some can send out the electrical signals referred to above in addition to being passive receivers. Most of them can be steered to point to different parts of the sky. Their relevance here is that it is quite feasible to send out signals which may be reflected by Comet Halley. Reflections have been obtained from the planet Venus, whose order of distance from the Earth can be the same as that of Comet Halley.

Two possibilities have been studied: (1) to use the steerable telescope of the Deep Space Network at Gold-stone, California, which, with its 210-foot-diameter dish, can pick up faint signals equivalent to one thousand-millionth of a watt in power. Opportunities for study of Halley's comet could exist for about 20 hours in April 1986; (2) to use the non-steerable but very much larger telescope at Arecibo in Puerto Rico. This was constructed by excavating a parabolic limestone bowl of 1000 foot diameter in the Earth and lining it with radio-wave-reflecting mesh. It is very sensitive, but it is dependent upon Halley's comet passing across its beam of operation as the Earth rotates on its daily motion. The possibilities for observation by this telescope would be limited to about 5 hours in November 1985.

Two other factors will affect success at either of these telescopes: (1) the reflecting properties of Comet Halley to radio waves are not known, but the assumption has been made that only one-tenth of the geometrical cross-section of the comet will be useful for radar reflection; (2) each radio-telescope generates signals known as 'noise' in its electrical circuitry. If the signals received back from Halley's comet are of less strength than the 'noise' the chances of detection are slim, though there are techniques to assist recovery of such a signal. For each telescope the estimated ratio of signal to noise is only about 3. The likelihood of success is thought to be slight.

Two further applications might also be investigated. Does the comet radiate in its own right? Radio telescopes can detect radiation from hydrogen. If any radiation from another celestial source has to pass through the long tail of the comet, can the radio-telescopes detect any refraction, or bending, of the radiation?

30   Comet Halley space probes – Giotto, Planet A and Venera

Like other unmanned space probes which have been sent to the planets, each of the five spacecraft being sent to Halley's comet is essentially an assemblage of instruments designed to collect data and transmit it back to Earth.

This however is too basic a view, for the 'payload', as it is called, has to be supported by systems to provide the power for the instruments and for communication with Earth. The communication has to be two-way so that commands can be given to the space probe as well as information received from it. The commands may operate further systems which will re-orientate the spacecraft or modify its trajectory. Additional systems are needed to protect the instruments from overheating and radiation and, particularly relevant to the Comet Halley space probes, from damage from the impact of dust from the comet and the casual small meteor.

The spacecraft are not driven by rocket motors in their journey from the vicinity of Earth to the vicinity of Comet Halley. Apart from the use of small control jets which can alter their attitude and give some manoeuvrability, the trajectory of each craft is dependent only on the gravitational attraction of the Sun, as indeed is that of Halley's comet itself. Nevertheless each spacecraft must be lifted from the Earth's surface and given sufficient velocity in the right direction, by the use of a large rocket. The Japanese will launch both their space probes directly from a Mu-3SII rocket. The European Space Agency will adopt an alternative method. An Ariane 3 rocket will first place its space probe Giotto into an orbit round the Earth. A few days later the spacecraft's own rocket motor will be fired and Giotto will be injected into its heliocentric trajectory to encounter Halley's comet on 13 March 1986.

Giotto

This is the name given to the spacecraft which the European Space Agency will send to within a few hundred kilometres of the centre of the coma of Halley's comet. The name is that of a painter, Giotto di Bondone, whose scene *The Adoration of the Magi* in the fresco in the Scrovegni Chapel in Padua is now famous (see plate 6). Giotto remembered the return of Halley's comet in AD 1301 and included a realistic detailing of the comet in his painting which he completed in 1304. The possible identification with the star of Bethlehem is mistaken, but Giotto's name will go down in history for a reason of which he could not even have dreamt.

Giotto's launch is planned for some time in the 15 days around 10 July 1985. This flexibility is necessary to facilitate allowances for the latest forecasts of the position of Halley's comet. If this so-called 'launch window' is not used there is no possibility of a rendezvous with Halley's

comet at this apparition. The launching site of the
spacecraft will be at Kourou in French Guiana in South
America. We ourselves can detect no motion at the surface
of the Earth but, due to the rotation of the Earth on its axis,
objects near the equator have a speed of about 1000 miles
per hour. Kourou is only about 4 degrees north of the
equator; so, since the spacecraft will be launched in the
direction of rotation of the Earth, it will receive maximum
assistance in achieving the velocity required for its 'parking' orbit. A few days later, when Giotto is nearest the
Earth in the parking orbit, its own solid propellant rocket
motor will be fired just as long as is necessary to transfer the
space probe into its trajectory to Halley's comet, increasing
its velocity by 1.4 km/s or 3150 miles per hour.

The paths of Giotto, the Earth and Comet Halley round
the Sun are represented in figure 25. Although Giotto will
be equipped with means to alter course slightly according
to the latest available information, it is expensive in fuel to
manoeuvre the craft out of the plane of its orbit, and so its
encounter with Halley's comet is planned to be very near
the plane of the Earth's orbit. Rendezvous will be on
13 March 1986.

So much for how Giotto will get near to Halley's comet;
what is it programmed to do when it gets there? We have
said that no-one knows whether a comet has a solid
nucleus, or whether the nucleus is merely an optical effect.
As long ago as 1971 Raymond Lyttleton suggested that the
only real means of testing this was to send a probe to a
comet. Giotto will, if all goes well, pass within 500 km of
such a nucleus (if it exists!).

Giotto is equipped with a multi-colour camera which, it
is hoped, will provide numerous images of the comet
nucleus with a resolution down to 30 metres. From these
the size, mass and rotation of the nucleus may be deduced.
Additionally, the camera images should yield the reflectiv-
ity and the shape of the comet, with any variation in the

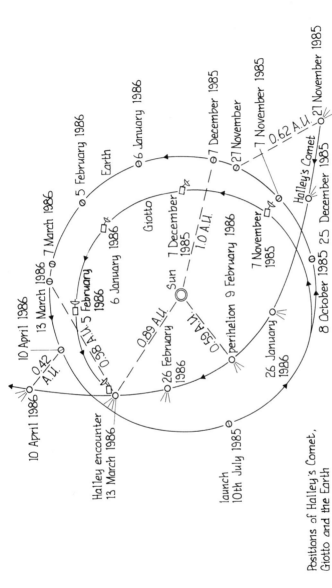

FIGURE 25   The path of Giotto in relation to Halley's comet, the Earth and the Sun.

wavelength of the light coming from the comet which will give information on its composition. The images will be reflected through 90° by a mirror into the camera so that the latter does not have to point directly at the oncoming dust particles. The camera can swivel forwards towards the comet, sweep the region of the coma, and point backwards towards the comet after the encounter. Giotto will be stabilized by rotating at 15 revolutions per minute at its encounter with Halley's comet, and this will provide the means for scanning its image across an array of charge-coupled detectors (see section 31) behind colour filters.

Apart from photographing the comet, the objectives of the Giotto mission are:

(a) to determine the elements and isotopes (section 18) of the volatile components of the coma and to determine the molecules from which they are derived;
(b) to do the same for the cometary dust particles;
(c) to measure the rate of production of gas and dust;
(d) to investigate what effects result from the interaction of the cometary plasma and the plasma of the solar wind.

The uncertainty about the structure of comets might also be cleared up (see section 18).

To carry out these tasks Giotto will be equipped to perform more than ten experiments. The technical details of these and the very sophisticated equipment carried are too lengthy and involved for a book of this nature, but it may be interesting to list the equipment here with the names of the countries responsible for its design. Some items of apparatus perform multiple tasks.

Multicolour camera	West Germany
Neutral mass spectrometer	West Germany
Ion mass spectrometer	Switzerland

Dust mass spectrometer	West Germany
Dust impact detector system	United Kingdom
Fast ion sensor	United Kingdom/France
Implanted ion sensor	United Kingdom/France
Electron electrostatic analyser	United Kingdom/France
Positive ion cluster composition analyzer	United Kingdom/France
Energetic particle monitor	Eire
Magnetometer	West Germany
Optical probe	France

The total electrical power required to drive all these instruments is about 50 watts, about that required for a desk light. More power will be needed for the communications systems. It will be derived from the array of solar cells which surround Giotto, which will receive maximum energy from the Sun because of the spacecraft's attitude and spin.

Data will be transmitted to Earth from a large dish antenna 1.5 metres in diameter to the 64-metre-diameter radio-telescope at Parkes in Australia. The antenna will be de-spun so that it can always point towards the Earth.

Although some of the experiments may be reset from Earth during Giotto's approach to the comet, there will be no possibility of making adjustments during its brief passage *through* the comet. The two-way time for any information and control message at encounter will be over 15 minutes.

The passage of Giotto through the comet, we hope through the coma, will mean impact from dust particles at high velocity (see section 9). A specially designed dust shield has been incorporated both to protect the instrumentation, all of which will lie behind the dust bumper, and to act as a dust sensor for a range of particles from 10^{-10} g to 0.1 g. This will be similar in form to that shown in figure 26, an annulus of nearly 2 metres outside diameter surrounding the central main rocket of Giotto. It is expected that the front sheet of the bumper shield will be eroded but that the protection provided overall will be sufficient for Giotto to come safely within 500 km of the nucleus.

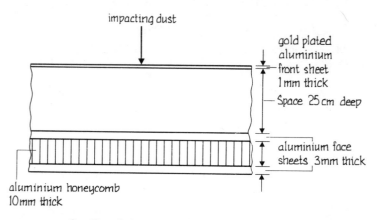

FIGURE 26   *Giotto's bumper shield for protection against impacting comet dust.*

The motion of Giotto is *direct*; that is, its trajectory takes it round the Sun in the same direction as the Earth's motion. The motion of Halley's comet is *retrograde* so that the relative speed of approach of the two principal characters in this venture will be about 68 km/s (153 000 miles per hour). This poses special problems, not merely because of the very short time during which the experiments can operate but because no one has any experience of the behaviour of microparticles as such a speed. At this speed each hydrogen atom in the comet has an energy which readily disinguishes it from the hydrogen atoms given off by the spacecraft by outgassing, so this large velocity of approach will assist the operation of the mass spectrometer. Ingenuity makes a virtue of necessity.

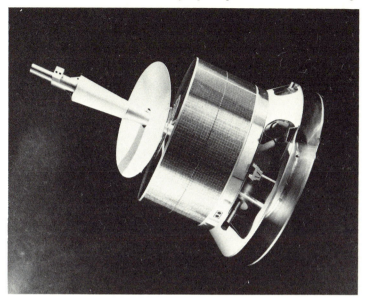

PLATE 17   *A model of the Giotto spacecraft showing the camera and some of the experimental equipment. The large de-spun antenna points permanently at the Earth. The bumper shield on the right protects the spacecraft from destruction by cometary dust particle impact.*

A model of Giotto showing the location of some of its components is shown in plate 17. Giotto at launch will have a mass of 950 kg and at encounter 512 kg, of which only 52 kg is payload. For detailed accounts of the equipment and experiments aboard Giotto the reader should refer to the *Yearbook of Astronomy 1983*, edited by Patrick Moore, and *Giotto Mission*, published by the European Space Agency in 1981.

Planet A

Although the suggestion may be implied in the name of Japan's space probe to Halley's comet, there is no direct connection with the search for the elusive tenth planet of the solar system, the so-called Planet X (see p. ooo). In the first case, Halley's comet is being investigated in a practical way, while the second case was a theoretical study of the motion of Halley's comet at previous apparitions in an attempt to discover a possible unknown planet in our solar system.

Planet A will be launched towards Halley's comet on an MU-3SII rocket, as will a second space probe, MS-T5. Both spacecraft will be launched directly into their heliocentric trajectories without using parking orbits around the Earth. The launch of Planet A from Japan's Institute of Space and Astronomical Science (ISAS) Space Centre at Uchinoura, Kagoshima, will take place during a second 20-day launch-window in late August of early September 1985. In choosing the exact time of launch the Japanese have to take into account not only astronomical considerations but also much mundane matters as the clearing of the sea of fishermen for their safety and the possibility that a typhoon might arise.

The orbit of the MS-T5 has yet to be decided but it may be directed specifically into Comet Halley's tail. Ion temperature, ion velocity, plasma density and electron temperature will be monitored.

Planet A will approach the nucleus of Halley's comet no closer than 100 000 km. Its main scientific objective is to make ultra-violet observations of the comet soon after it has passed perihelion, when activity is expected to be at a peak. The space probe's ultra-violet camera will take a series of pictures of the growth and decay of the hydrogen cloud around the comet nucleus (see section 18). Hydrogen radiates in the ultra-violet part of the spectrum (see section 23). As with the Giotto spacecraft, Planet A will be spin-stabilized, but at a lower rate of 5 revolutions per minute, reduced to one-half of a revolution per minute when scanning by the camera across charge-coupled devices is taking place. The three-dimensional distribution of the solar wind will be investigated by charged particle detectors. The total payload of Planet A has a mass of only 10 kg (plate 18).

Information will be transmitted from each spacecraft to a large dish aerial about 170 miles from Tokyo.

The nearest encounter of Planet A and Halley's comet will be on 7 March 1986 and that of MS-T5 one day later, just before Giotto's closest encounter on 13 March 1986, but of course these spacecraft will be far apart.

Venera–Halley Mission

This represents the contribution of the USSR, with the aim of learning more about Halley's comet and thereby of other comets, too. Information about this mission is not as readily available as for the European Space Agency and

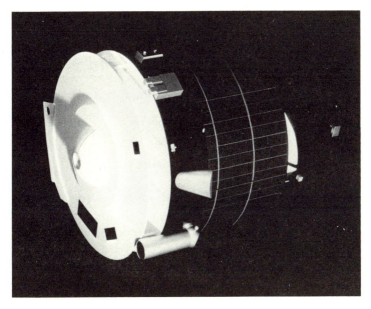

PLATE 18   *The main features of the Planet A space probe.*

Japanese missions, but it is known that the USSR will launch two Venera-type spacecraft in December 1984. They are numbers 17, 18 in the Venera series but will carry the special name Vega-Ve for Venus and ga for Halley, since the Russian alphabet includes no letter h. This is much earlier than Giotto and Planet A because both Venera craft will first carry entry probes to the planet Venus. Having delivered their experimental packages on Venus, they will use the gravitational effects of that planet to modify their trajectory to take them close to Halley's comet around 6–9 March 1986. The first Venera is targetted to encounter the comet at a distance of 10 000 km while the second, using positional data from the first, may come closer at 3000 km.

The payload of the Venera will have a mass of 120 kg, which is large compared with Giotto and Planet A. The spacecraft will have about 3½ hours to collect data. The experimentation of the Venera is very similar to that of Giotto, with a wide angle camera, a narrow angle camera and two spectrometers covering different bands.

All encounters of Giotto, Planet A and Venera will take place at roughly the same speeds relative to Halley's comet, about 70 km/s.

31   CCD and all that

Electronic devices are becoming more and more the tools of the astronomer. One of the most significant of these devices, the radio telescope, has been used extensively since 1932 when Karl Jansky first received the radio signals which come to us from the Milky Way, the faint band of stars which spreads across our sky. His receiving aerial was crude, mounted on Ford motor wheels for easy rotation. The amplifier in his receiving equipment converted the signals into a steady hissing sound in a loudspeaker.

Today aerials, or antennae, are much more sophisticated. The type most easily recognizable has a large dish, parabolic in shape, with a small aerial supported at the focus. There are literally thousands around the world. Although designed primarily to receive radio signals, they can also be used to transmit signals with directional precision. The associated electronics are also much more complex than those of Karl Jansky.

These aerials have a two-fold application in connection with Halley's comet (see section 29). The 210-foot diameter steerable dish at Goldstone, California, will be used to transmit signals towards Halley's comet and receive the echo reflected from it. The much larger 300-metre-diameter dish hollowed out of the ground at Arecibo, Puerto Rico, will be used similarly, but since it is not steerable its success will depend on Comet Halley passing across its operating arc as the Earth rotates. Two further questions are important and will be investigated using these aerials: does any part of the comet radiate its own signals, and are radio signals from the sky behind the comet refracted on passing through its tail?

The other main use for the aerials lies in the communicating systems between the Earth and the various space probes Giotto, Planet A and Venera described in the previous section. Communication from them will be vital, so that the information they gather will not be lost, since none of the spacecraft are expected to return. Signals can be sent *to* the spacecraft on their journey to the Halley encounter to modify their trajectories according to the latest known positions of the comet. Remember that Giotto is planned to approach the nucleus at 500 km or less at a distance from the Earth of some 147 million km. Even the programming of the experiments aboard Giotto can be altered from Earth by the communicating system.

Radio astronomy is a specialist science in itself, but the optical astronomers who use the more familiar type of telescope have not been slow to take full advantage of the latest developments in electronics. Large optical telescopes bring the light rays from objects in the sky to a focus where they are viewed through an eyepiece. A camera placed at the focal plane can then photograph these objects. Except for the Moon the light levels of planets and stars are very low, and long time exposures are often necessary for photography on normal photographic plate or film. If, however, instead of a photographic plate a device which converts light into some electrical quantity such as voltage or charge is placed in the focal plane, the image can be amplified and produced electronically. Such a device might be the *photomultiplier*, or a row of photomultipliers, in which the light falls on a photocathode, thus releasing electrons.

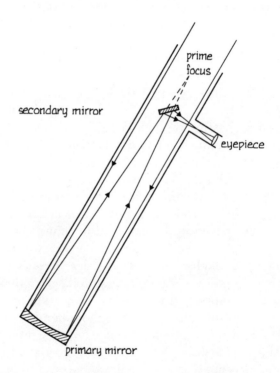

FIGURE 27   *A Newtonian reflecting telescope showing the prime focus.*

The electrons can be accelerated within the photomultiplier tube to strike successive dynodes, releasing progressively more and more electrons which are collected at the base of the tube. The *image converter tube* operates in the same way except that its physical arrangement is different and the final electrons strike a phosphor screen with a distribution of electrons similar to the distribution of light entering the telescope. The screen can be scanned just as the picture in a television set is scanned. The intensities of the parts of the image can be fed into the memory addresses of a computer, so that a continually intensifying picture can be reproduced at any time, even while the image is being built up.

The *charge-coupled device* has been mentioned twice before in this book. Halley's comet was found again on its present return by astronomers at Mount Palomar using a charge-coupled device (CCD) at the prime focus (see figure 27) of the large 200-inch optical telescope. The appearance of the comet at that time was very faint indeed and its distance very great (plate 12). The CCD, invented in 1969, is a physically small silicon 'chip' containing a large array of individual sensors (800 × 800 to obtain plate 12). The CCD consists of a three-layered n-type or p-type metal oxide semiconductor device (figure 28). There is one layer of metallic electrodes, often aluminium, and another of silicon crystal, separated by an insulating layer of silicon dioxide. When light from an image is focused on the CCD, a pattern of electrical charges is created. The charges vary in proportion to the amount of light and thus serve as an electrical representation of picture elements. The array can be scanned as described above and the device can store and transfer electrical charge signals at great speed. It is therefore eminently suitable for use with microcomputers. Examination of plate 12 will show that it is far from the conventional photograph to which we have been accustomed in the past. It is made up of small squares or 'pixels' arising from the scanning process. Any part of the photo-

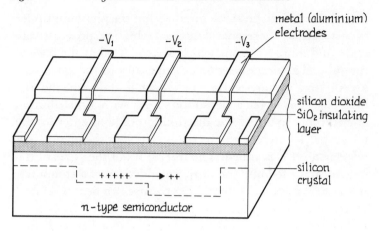

FIGURE 28   *A charge-coupled device (CCD) in the transfer mode.*

graph can be enhanced. The Giotto and Planet A space probes each carry CCDs. The spin of each spacecraft provides the means of scanning. The Venera craft are three-axis stabilized and do not normally spin.

Another use of electronics in the service of astronomy is the *microdensitometer.* A single conventional photographic plate taken by a large optical telescope can contain as many as a million images of stars and galaxies. The measuring of the brightness of each image on such a plate is difficult and tedious. In the microdensitometer the plate is scanned with a spot of light and the light is collected on a photoelectric cell after passing through the plate. The output from the cell is digitized (converted to a number) and passed to a computer which can then produce a picture such as the one of Halley's comet in plate 19. Four conventional photographic plates taken in 1910 were used by Perkin-Elmer Inc. in the United States to produce the picture (plate 19) in this manner. The contours, the *isophotes,* show the points which have the same brightness, just as the contours on an Ordnance Survey map show points of equal elevation above sea level.

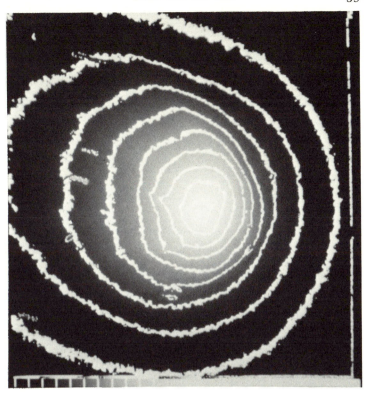

PLATE *19   D. A. Klinglesmith of NASA/Goddard Space Flight Center, Laboratory for Astronomy and Solar Physics, digitized the original Helwan plates and combined four original images into one. The enhanced image has been smoothed with a five by five box filter and every 10 signal level turn on to produce the contour levels.*

Very large telescopes, both optical and radio, are nowadays set on their sky targets by electrical/electronic control and are guided similarly to compensate for the effect of the rotation of the Earth.

VI

Conclusion

32 Halley's comet in the future

It is natural for us first to take a look at previous appearances of Halley's comet because most of what we know of the comet comes from past observations. What then, we may ask, does the future hold for Halley's comet? To answer this accurately we would need to be clairvoyant, but we may deduce some information by looking at the past motion of Halley's comet, and at the fate of other periodic comets.

It is possible, using principles of mechanics established by Newton and modern computing techniques, to predict accurately, for many comets, when they will be at perihelion on their next return. The gravitational effects of the planets, particularly of Jupiter, on the comet orbits can be included in these computations.

When this is done for Halley's comet for past appearances, the times forecast for perihelion passage fail to agree with reality by three or four days. Joseph L. Brady of the University of California has overcome this problem mathematically, but only by including a *non-gravitational* force in his equations. The nature of this force is not known, and to this extent the future of the orbit is uncertain. The loss of material from the comet and the forces which this produces as the comet approaches the Sun every 76 years must at least make a contribution to the non-gravitational force, and as such it is an indeterminable quantity.

Sir Fred Hoyle suggests that the general effect of the gravitational forces produced by the planets on all periodic comets is to make their elliptical orbits more round. There was a close approach of Halley's comet to Jupiter in 1838 and there have been at least six close approaches to the Earth–Moon system and four to Venus since 918 BC. The length of the orbit of Halley's comet, now stretching out to the orbit of Pluto, will eventually become much shorter. When most of Comet Halley's volatile material has been lost this comet may be orbiting the Sun into the region between the orbits of Mars and Jupiter, which Hoyle graphically describes as the 'graveyard of comets'. Halley's comet is not a small one, however, and the erosion of material will take a long time. There is of course the possibility that any comet approaching the Sun will hit the Earth. What an ignominious end for such a famous comet as P/Halley, and most likely the same for the Earth! But Sir Fred Hoyle puts the chances of any comet hitting the Earth as one in a billion, or in English terminology one in 1000 million; if the comet breaks up, and some comets *have* been seen to divide, the likelihood of impact is put at about once every 10 million years. The nearest approach of Halley's comet known so far is 3.7 million miles, although Halley himself was very much concerned with the possibility of impact.

We may predict with some confidence that if we ourselves do not see Halley's comet at its next return, perihelion passage already calculated as 29 July 2061, some of our children or our children's children will. When Halley's comet last appeared, the hansom cab was a convenient method of transport. Today we have the fast motor car, the supersonic aeroplane, the space vehicles. With the rapid development of modern high technology, who can say what will be available at P/Halley's next appearance? We shall almost certainly know much more about comets than we do today.

APPENDIX A
Notes on the microcomputer programs

Four programs relevant to the text of this book have been provided in sections 5, 11, 15, 25. A version of each program has been given for each of three common micro-computers, namely the ZX81 (given in appendix B), the Spectrum and BBC Model B, apart from ALTAZ, which has been written for the BBC Model B and the Spectrum only. The programs are all in BASIC. They have been so structured that they can easily be adapted, with little or no alteration, for other makes of microcomputer. All the programs have in-built prompts and a suitable value has been provided for each input. On the ZX81, all programs will need to use the 16K memory.

KEPLER *(section 5)*

Halley's comet moves much more quickly when it is near the Sun than when it is in the remote regions near the orbits of Neptune and Pluto (see section 9). This program gives a visual impression of the motion of a periodic comet for one complete revolution round the Sun, by plotting its positions at equally spaced intervals of time. The parameters of the orbit of Halley's comet do not readily lend themselves to such a demonstration. In particular the eccentricity of its elliptical orbit (0.967) is too close to unity. Consequently the parameters of the orbit demonstrated have been chosen to give a realistic impression of the movement, but Halley's comet obeys the same laws.

The position at any time depends upon the determination of an angle, the *eccentric anomaly E*, from the transcendental equation

$$n(t - T) = E - e.\sin E$$

where t is the time, T the time at perihelion, n the mean motion and e the eccentricity of the orbit. This is *Kepler's equation*.

In the program Newton's method has been applied to find E and as this is a method of successive approximation the calculations for 24 dates take some time. The ZX81 takes about 1½ minutes before the display begins. The other computers take much less time. All the results are obtained before any display, so it is possible to plot the positions at truly equal time intervals.

Advantage has been taken of the facilities of the BBC model and of the Spectrum to extend this program to illustrate Kepler's second law of planetary motion, namely that the line joining the Sun to the comet sweeps out equal areas in equal times. The areas displayed have been calculated on the basis of triangles as opposed to the sectors of the ellipse and some error can be expected for those areas nearest the Sun. (Kepler stated two other laws which are worthy of investigation.)

ALTAZ *(section 11)*

This program calculates the altitude (degrees above the horizon) and the azimuth (bearing in degrees measured eastwards from North) of Halley's comet at any time of day or night on any date in 1985 or 1986 for any observer at any latitude and longitude. For those readers who are familiar with astronomical coordinates the program also gives the corresponding right ascension and declination of the comet (see section 11).

This period extends beyond the dates when Halley's comet should be visible to the naked eye and the program will therefore be useful in locating the comet with binoculars or telescope. It is accurate to about one degree in each coordinate.

The program calculates, additionally, the coordinates of the Sun and indicates whether or not the sky is dark at the time selected. The criterion for a dark sky in this context is that the Sun should be more than 18 degrees below the horizon (see section 13).

The program does not rely on the use of any astronomical tables. It is written in BASIC in such a way that it will run directly on the BBC and Spectrum computers and similar models, and with very little modification on the ZX81 computer with 16K RAM.

At the appropriate prompt in the program the reader is asked to enter his latitude in degrees, positive for north, negative for south, the year, the month (as a number), the date, and the time (hours and minutes) reckoned on the 24-hour clock.

The program is based on the methods described in Dr Peter Duffet-Smith's *Practical Astronomy with your calculator* (Cambridge University Press, 1979).

MOON *(section 15)*

Halley's comet will be seen more easily when the sky is not brightened by the presence of a full Moon. This program asks you to input the year, most likely 1985 or 1986 AD, and the calendar, which for this apparition of Halley's comet will be Gregorian G (see section 13). The month is entered into the computer as a number according to the instruction in each program. If you require the date of the full moon at your specified month you should enter 15 at the appropriate prompt, zero for new moon or even, for example, 7 for a 7-day-old moon.

The program is empirically based. It has no matching theory, but the basis of it is described in 'A thumbnail almanac for the Moon' in *Sky and Telescope*, June 1974.

HALLEY *(section 25)*

When Halley's comet is near the Sun the difference between its elliptical orbit and a parabolic orbit is not very great. Indeed Newton devised his method of determining the parameters of the orbits of comets on the assumption that the orbits were parabolic, a method which Halley himself used.

This program has been devised on this assumption and also assumes that the orbit of the Earth round the Sun is circular, which it very nearly is. The program illustrates the positions of Halley's comet and the Earth relative to the Sun at 13 dates of this apparition. The dates chosen begin with the launching of Giotto in 1985 July and end one year later. The comet is represented by ' , the Earth by · and the Sun by +.

The position of the comet at any time is defined by the *true anomaly* v. This angle must be determined from the cubic equation

$$\frac{1}{3} \tan^3 \frac{v}{2} + \tan \frac{v}{2} = k \ (t - T)/(\sqrt{2}.q^{3/2})$$

where k is a constant known as the Gaussian constant, t is the time, T the time of perihelion passage and q the perihelion distance.

In the program Newton's method of successive approximation has again been used to determine v for 13 dates. All the calculations are completed before the display so the position of the comet can be shown at truly equal time-intervals of about one month for one year.

APPENDIX B
Computer programs for the ZX81

Kepler

```
10 REM "KEPLER"
20 PRINT "ORBIT DEMONSTRATION"
30 PRINT "ENTER SEMIMAJOR AXIS (1-20)"
40 INPUT A
50 IF A<1 OR A>20 THEN GOTO 30
60 PRINT "ENTER ECCENTRICITY (0-1)"
70 INPUT E
80 IF E<0 OR E>0.95 THEN GOTO 60
90 PRINT "ENTER TIME OF PERIHELION PASSAGE (YEAR)"
100 INPUT PP
110 PRINT "ENTER YEARLY MOTION (DEGREES PER YEAR)"
120 INPUT YM
130 PRINT "ENTER INITIAL TIME (YEAR)"
140 INPUT ST
150 PRINT "ENTER NUMBER OF DATES (10-100)"
160 INPUT U
170 IF U<10 OR U>100 THEN GOTO 150
180 PRINT "ENTER TIME INTERVAL (YEARS E.G. 0.5)"
190 INPUT TI
200 CLS
210 PRINT "PLEASE WAIT..."
220 DIM X(100)
230 DIM Y(100)
240 LET B=A*SQR (1-E*E)
250 LET Q=A*(1-E)
260 LET RAD=PI/180
270 LET THETAE=0
280 FOR I=1 TO U
290 LET THETAC=YM*(ST+(I-1)*TI-PP)*RAD
300 REM NEXT APPROXIMATION
310 LET DIFF=THETAC-(THETAE-E*SIN THETAE)
320 LET THETAE=THETAE+DIFF/(1-E*COS THETAE)
330 IF ABS (DIFF)>0.0001 THEN GOTO 300
340 REM CALCULATE CARTESIAN COORDINATES
350 LET X(I)=A*(COS THETAE-E)
360 LET Y(I)=B*SIN THETAE
370 NEXT I
380 REM NOW TRANSLATE SO THAT ORBIT IS CENTRED ON SCREEN
390 FOR I=1 TO U
400 LET X(I)=INT(30.5-Q+X(I))
410 LET Y(I)=INT(11.5+Y(I))
420 NEXT I
430 FOR J=0 TO 21
440 FOR I=0 TO 31
450 PRINT AT J,I; CHR$ 128
460 NEXT I
470 NEXT J
480 PRINT AT 11,30-INT(Q+0.5); CHR$ 151
490 FOR I=1 TO U
500 IF Y(I)>=0 AND Y(I)<=21 AND X(I)>=0 AND X(I)<=31 THEN
    PRINT AT Y(I),X(I); CHR$ 154
510 PAUSE 50
520 NEXT I
```

Moon

```
10 REM "MOON"
20 PRINT "MOON PHASE"
30 DIM R(19)
40 DIM L(33)
50 DIM M(12)
60 PRINT "INPUT R"
70 PRINT "ENTER 9,25,10,26,12,28,
   13,29,15,1,17,3,19,4,20,6,22,7,23"
80 PRINT
90 FOR I = 1 TO 19
100 INPUT R(I)
110 PRINT R(I);","
120 NEXT I
130 STOP
140 CLS
150 PRINT "INPUT L"
160 PRINT "9,28,17,6,25,14,3,22,
   11,30,19,8,27,16,5,24,13,2,21,10,
   29,18,7,26,15,4,23,12,1,20,9,28,17"
170 PRINT
180 FOR I = 1 TO 33
190 INPUT L(I)
200 PRINT L(I);","
210 NEXT I
220 STOP
230 CLS
240 PRINT "INPUT M"
250 PRINT "ENTER 0,2,2,4,4,6,7,8,9,10,11,13"
260 PRINT
270 FOR I=1 TO 12
280 INPUT M(I)
290 PRINT M(I);","
300 NEXT I
310 STOP
320 CLS
330 PRINT "ENTER YEAR "
340 INPUT Y
350 LET YR=Y
360 IF Y<0 THEN GO TO 330
370 PRINT "BC OR AD"
380 INPUT A$
390 IF A$<>"BC" AND A$<>"AD" THEN GO TO 370
400 PRINT "ENTER MONTH NUMBER"
410 PRINT "11=JAN, 12=FEB, 1=MAR,...,10=DEC"
420 INPUT Z
430 IF Z<1 OR Z>12 THEN GO TO 400
440 PRINT "ENTER J FOR JULIAN OR G FOR GREGORIAN"
450 INPUT B$
460 IF B$<>"J" AND B$<>"G" THEN GO TO 440
470 PRINT "ENTER PHASE (0 FOR NEW UP TO 15 FOR FULL)"
480 INPUT C
490 IF C<0 OR C>29 THEN GO TO 470
500 CLS
510 IF Z>10 THEN LET Y =Y-1
520 IF A$="AD" THEN LET J =Y+4712
530 IF A$="BC" THEN LET J=4713-Y
540 LET Q=INT (J/76)
550 LET N=J-Q*76
560 LET S=INT (Q/4)
570 LET R= INT (N/4)
580 LET L =N-R*4
590 FOR I = 1 TO 30
600 IF L(I)=R(R+1) THEN LET D=L(I+L)
610 NEXT I
620 LET W=D-S-M(Z)+C
630 IF B$="J" THEN LET T=W
640 IF Y>1582 AND Y<=1700 THEN LET Y=10
650 IF Y>1700 AND Y<=1800 THEN LET Y=11
660 IF Y>1800 AND Y<=1900 THEN LET Y=12
670 IF Y>1900 THEN LET Y=13
680 IF B$="G" THEN LET T=W+Y
```

```
690 IF T<0 THEN LET T=T+30
700 IF T>30 THEN LET T=T-30
710 PRINT
720 PRINT "A MOON ";C;" DAYS OLD OCCURS ON"
730 PRINT "DAY ";T;", MONTH ";Z;", YEAR ";YR;" ";A$
740 GO TO 330
```

Halley

```
10 REM "HALLEY"
20 PRINT "HALLEY,EARTH,SUN DEMO"
30 PRINT
40 DIM A(40)
50 DIM B(40)
60 DIM C(40)
70 DIM D(40)
80 DIM H(40)
90 DIM M(40)
100 DIM N(40)
110 DIM R(40)
120 DIM X(40)
130 DIM Y(40)
140 PRINT "ENTER GAUSSIAN CONSTANT (AU/YEAR) "
150 INPUT K
160 PRINT "ENTER PERIHELION DISTANCE (AU) "
170 INPUT Q
180 PRINT
190 PRINT "ENTER YEAR OF PERIHELION PASSAGE "
200 INPUT P
210 PRINT "ENTER MONTH OF PERIHELION PASSAGE "
220 INPUT M
230 IF M<1 OR M>12.99 THEN GO TO 210
240 LET P=INT(P)+(M-1)/12
250 PRINT "ENTER YEAR OF INITIAL DATE "
260 INPUT S
270 PRINT ENTER  "MONTH OF INITIAL DATE "
280 INPUT M
290 IF M<1 OR M>12.99 THEN GO TO 270
300 LET S=INT(S)+(M-1)/12
310 PRINT "ENTER TIME INTERVAL (MONTHS) "
320 INPUT W
330 LET W=W/12
340 PRINT "ENTER NUMBER OF DATES (5-40) "
350 INPUT U
360 IF U<5 OR U>40 THEN GO TO 340
370 PRINT "PLEASE WAIT..."
380 FOR I=1 TO U
390 LET M(I)=K*(S+(I-1)*W-P)/(SQR (2*Q**3))
400 IF I=1 THEN LET A(I)=M(I)
410 LET H(I)=-(A(I)**3+3*A(I)-3*M(I))/(3*A(I)*A(I)+3)
420 LET B(I)=A(I)+H(I)
430 LET C(I)=B(I)**3+3*B(I)-3*M(I)
440 LET D(I)=ABS C(I)
450 IF D(I)<0.0001 THEN GO TO 480
460 LET A(I)=B(I)
470 GO TO 410
480 LET N(I)=ATN B(I)
490 NEXT I
500 FOR I=1 TO U
510 LET R(I)=Q*(1+TAN N(I)*TAN N(I))
520 LET N(I)=2*N(I)
530 LET X(I)=4*R(I)*COS N(I)
540 LET Y(I)=4*R(I)*SIN N(I)
550 LET X(I)=20-Q+X(I)
560 LET Y(I)=10+Y(I)
570 NEXT I
580 PRINT AT 10,20; CHR$ 23
590 FOR I=1 TO U
600 PRINT AT INT(Y(I)+0.5),INT (X(I)+0.5); CHR$ 26
610 PRINT AT 10-4*(SIN ((I-2)*PI/6)),20+4*COS ((I-2)*PI/6)); CHR$ 21
620 PAUSE 50
630 NEXT I
```

APPENDIX C
How to report the discovery of a new comet, or the recovery of an old one

The discovery of a new comet, or the recovery of an old one may be reported to the Central Bureau of Astronomical Telegrams, Smithsonian Astrophysical Observatory, 60 Garden Street, Cambridge, Mass., 02138, USA, by sending a telegram. The telegram is recorded on a telex machine and should take a standard form which is self-checking against errors in transmission. It should be addressed to TWX 710 320 6842 Astrogram Cam.

As an example we give a telegram for the recovery of Comet Halley. The detail is not far from the reality.

JEWITT/DANIELSON   COMET
JEWITT/DANIELSON   19501   21016   12500
07110   20933   01252   82312   29295   JEWITT

Taking this example, reading the items in turn, we have the following explanation:

JEWITT/DANIELSON – the names of the discoverers

COMET – the type of object being reported. Other objects might be nova or supernova

JEWITT/DANIELSON – the names of the observers

19501 – the epoch (1950), which is the date of the star charts used to determine the comet's position, followed by '1' since the position given is approximate. It would be '2' for an accurate position, '3' for orbital elements or '4' for an ephemeris (a timetable)

21016 – the last digit of the year (2) of discovery, in this case 1982, the month (10) for October and the day (16)

12500 – the time of discovery of the above data expressed as a decimal part of the 24-hour clock: 12500 corresponding to 0300 hours or 3/24 = 0.12500 of a day

07110 – the right ascension of the comet (07 h 11.0 m)

20933 – the declination of the comet preceded by '1' for negative or south declination or '2' for positive or north declination corresponding to a declination of (+09° 33′)

01252 – the appearance of the comet: the first digit is always zero (0) for an approximate position; the second digit indicates either the total magnitude (1) or the magnitude of the nucleus (2) is being reported followed by the magnitude itself (25). The final digit in this group reports on whether the object is starlike or diffuse and gives an indication of the length of the tail (if any). Halley's comet showed no coma on recovery and has been accorded the digit '2'

82312 – as a check this group is the last five digits of the total of all the preceding groups

29295 – as a second check, this group is the last five digits of the sum of the groups giving the right ascension (07110), the declination (20933) and the information on the magnitude (01252)

JEWITT – the name of the sender of the telegram

This standard form of telegram allows Dr Brian Marsden at the Central Bureau to certify the discovery or recovery, as indeed he did with the recovery of Halley's comet.

Bibliography

Books

Apianus, Peter, *Astronomicum Caesareum*, 1504.

Armitage, A., *Edmond Halley*, Thomas Nelson, 1966.

Brandt, J.C., 'Comets', in *Readings from Scientific American 1909–1981*, 1981.

Calder, N., *The Comet is Coming*, BBC, 1980.

Hnatek, E.R., *A User's Handbook of Semi-conductor Memories*, John Wiley, 1977.

Hoyle, F., *Astronomy*, Macdonald Educational, 1962.

Hoyle, F. and Wickramasinghe, N.C., *Lifecloud*, Dent and Sons, 1978.

Hoyle, F. and Wickramasinghe, N.C., *Space Travellers*, University of Cardiff Press, 1981.

Hoyle, F. and Wickramasinghe, N.C., *Diseases from Space*, Dent and Sons, 1979.

Knight, D.L., *Johannes Kepler*, Chatto & Windus, 1962.

Larousse Encyclopedia of Astronomy, Hamlyn, 1959.

Lyttleton, R.A., *Mysteries of the Solar System*, Oxford University Press, 1968.

Maclagan, E., *The Bayeux Tapestry*, Penguin Books, 1949.

Moore, Patrick, *Sun, Myths and Men*, Frederick Muller, 1968.

Moore, Patrick, *Guide to Comets*, Lutterworth Press, 1977.

Moore, Patrick (ed.), *Yearbook of Astronomy 1983*, Sidgwick & Jackson, 1982.

Needham, J., *Science and Civilisation in China*, vol. 3, Cambridge University Press, 1959.

Newman, J.R., *The World of Mathematics*, Simon & Schuster, 1956.

Porter, J.G., *Comets and Meteor Streams*, Chapman & Hall, 1952.
Ridpath, Ian, *Signs of Life*, Penguin Books, 1977.
Ronan, C., *The Astronomers*, Evans Brothers, 1964.
Sandford, V., *A Short History of Mathematics*, Harrap.
Tattersfield, D., *Projects and Demonstrations in Astronomy*, Stanley Thornes, 1979.
Whipple, F.L., 'The nature of comets', in *Readings from Scientific American 1956–1975*, 1975.
Yeomans, D.K., *The Comet Halley Handbook*, NASA (JPL), 1981.

Journal articles

'Accounts of the Royal Astronomical Society', *Monthly Notices of the Royal Astronomical Society*, 4 (1836), p. 30.
Boyd, R.F.L., 'The Halley Lecture 1981', *The Observatory*, 101 (October 1981).
Brady, J.L., 'Halley's comet AD 1986 to BC 2647', *Journal of the British Astronomical Association*, 92, 5 (1982), p. 209.
'Drawings of Halley's comet', *Memoirs of the Royal Astronomical Society*, 10 (1835).
'Giotto and Halley's comet', *ESA Newsletter*, no. 4 (1982).
Gire, B., 'The Ariane launcher', *Spaceflight*, 24, 9 & 10 (1982), p. 358.
Haig, G.Y., 'A stellar spectrograph', *Journal of the British Astronomical Association*, 85, 5 (1975), p. 408.
Halley, E. 'Synopsis Astronomiae Cometicae', *Philosophical Transactions of the Royal Society, London*, 24, 297 (1705), p. 1882.
Ho Peng Yoke, 'Ancient and mediaeval observations of comets and novae in Chinese sources', *Vistas in Astronomy*, 5 (1962), p. 148.
'Japan's Halley comet probes', *Spaceflight*, 25, 6 (1983), p. 272.
Kiang, T., 'The past orbit of Halley's comet', *Memoirs of the Royal Astronomical Society*, 76 (1971), p. 43.
Lyttleton, R.A., 'Does a continuous solid nucleus exist in comets?' *Astrophysics and Space Science*, 15 (1972), p. 175.
McLaughlin, W.I., 'The natural history of Halley's comet', *Journal of the British Interplanetary Society*, 34, 7 (1981), p. 267.
'Periodic Comet Halley 1982i', *British Astronomical Association Circular*, no. 631 (1982).
Plummer, H.C., 'The Halley Lecture 1942', *Nature*, 150 (9 August 1942).

'Rendezvous with Halley's comet', *Spaceflight*, 24, 9 & 10 (1982), p. 369, and 25, 4 (1983), p. 157.

Ridley, H.B., 'Comets', Presidential Address, *Journal of the British Astronomical Association*, 83, 3 (1978), p. 227.

Shinn, B.F., 'Models of cometary orbits', *Journal of the British Astronomical Association*, 88, 2 (1978), p. 139.

Shove, D.J., 'Halley's Comet', *Journal of the British Astronomical Association*, 65, 7 (1955).

Whipple, F.L., 'The acceleration of Comet Encke', *Astrophysics Journal*, 111 (1950), p. 375.

Yeomans, D.K. and Kiang, T., 'The long-term motion of Comet Halley', *Monthly Notices of the Royal Astronomical Society*, 197 (1981), p. 644.

See also:

Journal of the British Astronomical Association: C. Ponnamperuma, 92, 4 (1982), p. 202, Review of 'Comets and the Origin of Life'; R. A. Lyttleton, 92, 4 (1982), p. 201, Review of 'Introduction to Comets'; and H. F. Michielson and J. L. Brady, 93, 4 (1983), p. 180, Letter: 'The Past Orbit of Halley's Comet'.

The Physics Teacher, November 1968, 'Laboratory Exercises in Astronomy'.

Newspapers and miscellaneous publications

The Astronomical Almanac 1983, HMSO, 1983.

Comet Bulletin, no. 19, British Astronomical Association, 1983.

'Futures', *The Guardian*, 30 October 1980.

Handbook of the British Astronomical Association 1983.

'The Night Sky', *The Guardian*, 30 December 1982.

Seeing Halley's Comet, International Halley Watch leaflet, 1983.

Giotto mission, European Space Agency, 1981.

Index